What Your Colleagues Are Saying . . .

"This book's title reveals its promise. *Classroom-Ready Rich Math Tasks* offers teachers an abundant collection of practical challenges for engaging their students in building mathematical fluency and developing strategic reasoning—exactly what all the Mathematics Standards expect."

Jay McTighe
Author and Educator
McTighe and Associates Consulting

"This book is exactly what every K–1 teacher needs! The 56 tasks are engaging, easily implemented, and most important, 'classroom ready.' The instructional strategies presented in the beginning chapters and the reflections emphasized in the last chapter provide the appropriate foundation for teachers of young children. As an experienced K–1 teacher, I guarantee that I would use the ideas presented and would encourage others to do so."

Juanita Copley
Professor Emerita
University of Houston

"As a district math leader, I appreciate this series because I know that any task my teachers choose will be high quality. I also appreciate the research-informed coaching provided in the first few chapters that will guide my teachers through selecting, planning, and implementing tasks to maximize their potential."

Brian Bushart
Elementary Mathematics Curriculum Coordinator
Round Rock Independent School District

"What a wonderfully practical and thought-provoking resource. Helpful for all of us, no matter where we are in our mathematics teaching journey. The shared tasks are enriched with strategies and tools for supporting student explaining and participation along with ways of understanding student thinking and the mathematics."

Megan Franke
Interim Chair and Professor of Education
UCLA

"The core of great math teaching is the high-quality task, brought to life through rich interactions. These tasks must be engaging and both mathematically and pedagogically generative—of interesting content *and* mathematical processes and practices. This valuable book is filled with such tasks, as well as guidance in planning and implementing them and materials for maximum enjoyment and benefit for teachers and their students."

Douglas H. Clements
Distinguished University Professor, Kennedy Endowed Chair in Early Childhood Learning, and
Executive Director of the Marsico Institute for Early Learning
University of Denver

"I highly recommend this book, which provides a research-based framework for engaging in rich tasks that are connected to and build on previous mathematical understanding, relevant contexts, and students' experiences and resources, as well as connecting to mathematics teaching practice. The high cognitively demanding tasks included support the development of important mathematics norms, practices, and concepts."

Robert Q. Berry, III
Samuel Braley Gray Professor of Mathematics Education
University of Virginia

"This book answers the often-asked question, 'Where can I find good tasks?' The authors have assembled a collection of rich mathematical tasks and provided guidance on how to plan and implement lessons around them in ways that will support opportunities for students to learn mathematics with understanding. This book is a game changer for kindergarten and first-grade teachers who want their students to do math!"

Margaret (Peg) Smith
Emeritus Faculty
University of Pittsburgh

"Finding the right task that invites young learners to think and reason mathematically isn't always easy. This impressive book offers various high-quality tasks and addresses planning and implementation considerations. It is an invaluable resource for everyone who strives to provide worthwhile learning experiences for kindergarten and first-grade mathematicians!"

Susie Katt
K–2 Mathematics Coordinator
Lincoln Public Schools

"By integrating the latest research with a powerful teacher voice, the authors have written a must-read for every teacher of mathematics who wants their students to have a more motivating and powerful classroom experience. But the authors don't simply argue that teachers include high-cognitive demand tasks in their teaching—they take the next steps and actually demonstrate *how* to plan for, implement, and support these tasks so each student can authentically engage in *doing* mathematics!"

Matt Larson
Past President
National Council of Teachers of Mathematics

"*Classroom-Ready Rich Math Tasks, Grades K–1* is an excellent resource for teachers, teacher educators, and professional development providers who want to engage students in high-level, cognitively demanding mathematics tasks. The book does an excellent job of explaining the why and the how of implementing doing mathematics tasks. To top it off, this book has a treasure trove of rich mathematics tasks for teachers to choose from."

Kyndall Brown
Executive Director
California Mathematics Project

"This book is designed explicitly to support teachers as they reflect on their math teaching practice either individually—or in collaboration with colleagues. Not only does it pull together the essential research connected to planning, setup, and implementation of high-cognitive-demand tasks, it also helps them to implement a full, ready-made collection of tasks, and then eventually move to selecting, adapting, and creating tasks on their own."

Nicole Rigelman
Professor of Mathematics Education
Portland State University

Classroom-Ready Rich Math Tasks, Grades K–1:
The Book at a Glance

Each task includes a short description of the mathematical idea that will be tackled as well as extensive preparation and launch notes.

Grade 1

Mathematics Standards

- Solve word problems that call for addition of three whole numbers whose sum is less than or equal to 20, e.g., by using objects, drawings, and equations with a symbol for the unknown number to represent the problem.

- Add and subtract within 20, demonstrating fluency for addition and subtraction within 10. Use strategies such as counting on; making ten (e.g., $8 + 6 = 8 + 2 + 4 = 10 + 4 = 14$); decomposing a number leading to a ten (e.g., $13 - 4 = 13 - 3 - 1 = 10 - 1 = 9$); using the relationship between addition and subtraction (e.g., knowing that $8 + 4 = 12$, one knows $12 - 8 = 4$); and creating equivalent but easier or known sums (e.g., adding $6 + 7$ by creating the known equivalent $6 + 6 + 1 = 12 + 1 = 13$).

Mathematical Practices

- Look for and make use of structure.
- Make sense of problems and persevere in solving them.

Task 16
Come to the Playground

Solve addition word problems with three addends

TASK

Come to the Playground!

Figure 8.3 The Playground

Source: robuart/iStock.com

There were 4 children on the playground. Soon, 5 more children came to the playground. Then 6 more children arrived at the playground. How many children are on the playground (Figure 8.3) now?

TASK PREPARATION

- This task involves the Add to, Result Unknown addition situation, with examples that facilitate students' use of making tens.

- Provide students with counting materials.

- Because there are three addends, you should be mindful about using the numbers that your students are comfortable with.

- Students should solve the task independently, and then compare equations and share solution strategies in pairs or in small groups.

- When writing problems with the Add to, Change Unknown situations, make sure that the word problem includes a specific action. Explicit actions help the student model the problem.

- Building off of Add to, Result Unknown situations, the inclusion of another addend allows students to explore the associative property and add the number of children on the playground in multiple ways.

- When the students experience problems with three addends, anticipate their solution strategies. Will they add the three numbers in the order that they are given in the problem (e.g., $4 + 5 + 6$), or, if they are becoming more flexible with their thinking, will they make a 10 with the 4 and the 6 and then add 5 to get 15.

Each task is labeled with the grade level and calls out the Mathematics Standards, Mathematical Practices, and Important Vocabulary highlighted in the task.

Each task includes a list of the necessary materials needed to complete it.

LAUNCH

1. Bring the class together as a whole group and have a discussion about situations that happen during recess and after lunch, when some children go to the playground first and then more children join them.

2. Then, either go to the playground or designate a spot in the classroom as the playground.

3. Say, "The math problem that we are going to start with today is this: There were 4 children on the playground. Soon, 5 more children came to the playground. Then 6 more children arrived at the playground. How many children are on the playground now?"

4. Say, "Let's act it out together! How many children start on the playground? (4) So I need four volunteers to come on to the playground. Then what happens in the story? Five more children come to the playground. So what should we do now? (Count out the five children that move over to the playground.) Then what happens? Six more children arrive at the playground! (Count out the six additional children that go to the playground). How can we find out how many children are on the playground now?" (Count out all of the children on the playground from 1 to 15.)

5. Return to the classroom and present the following problem on the board:

   There were ＿＿ children on the playground. Soon, ＿＿ more children came to the playground. Then ＿＿ more children arrived at the playground. How many children are on the playground now?

6. Read the problem to the class and say, "We solved a problem that had 4, 5, and 6 children at the playground. For this problem, pick three cards from your number cards. Those are the numbers of children that arrive at your playground."

ACCESS AND EQUITY

The focus of this task is to solve the problem, not to read it. Eliminate a potential barrier to task access by reading the problem and repeating it if needed.

FACILITATE

1. Make sure students have access to all of their mathematics tools, such as counters, ten frames, fingers, and things to draw with. Have the students solve the task in multiple ways if possible, sharing their strategies with at least one other student.

Vocabulary

- add
- add to
- combine
- join
- sum
- model
- equation
- numbers 1 to 20

Materials

- counters
- ten frames (optional)
- Number Cards student page (1–10, cut into sets), one set per student
- Come to the Playground student page, one per student

Extensive facilitation notes support educators in guiding the implementation of the task-lesson.

All activity sheets and online resources provided in the Materials section of each task are available for download or viewing on the companion website.

2. Distribute the The Triangles Have It recording sheet to student pairs. Allow student pairs to work on the task.

3. **Observe** the student pairs as they use trial and error to fill the shapes with triangles and figure out which shape uses more than six triangles. Also observe the students as they place the shapes on the template. What strategies do students use to fit the pattern block shapes on the template?

4. Facilitate a *Pair-to-Pair Share* with students. Allow time for students to share their thinking,

TASK 54: THE TRIANGLES HAVE IT STUDENT PAGES

online resources — To download printable resources for this task, visit **resources.corwin.com/ ClassroomReadyMath/K-1**

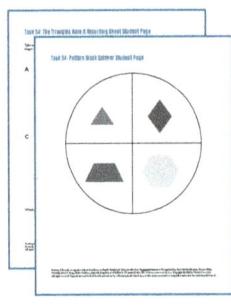

POST-TASK NOTES: REFLECTION & NEXT STEPS

Additional room for writing follows each task to allow for personalization as you plan the use of the lesson for your classroom.

The tasks in this book have been designed with student considerations in mind. Important ideas and teaching moves that will enrich students' learning experiences while engaged in *doing-math tasks* can be found in marginal boxes throughout the tasks. These ideas highlight access and equity, productive struggle, alternate learning environments, and strengths spotting.

Classroom-Ready
RICH MATH TASKS

GRADES
K–1

Classroom-Ready
RICH MATH TASKS

GRADES
K–1

Engaging Students in Doing Math

Beth McCord **KOBETT** • Francis (Skip) **FENNELL** • Karen S. **KARP**
Delise **ANDREWS** • Latrenda **KNIGHTEN** • Jeff **SHIH**

CORWIN Mathematics

FOR INFORMATION:

Corwin
A SAGE Company
2455 Teller Road
Thousand Oaks, California 91320
(800) 233–9936
www.corwin.com

SAGE Publications Ltd.
1 Oliver's Yard
55 City Road
London, EC1Y 1SP
United Kingdom

SAGE Publications India Pvt. Ltd.
B 1/I 1 Mohan Cooperative Industrial Area
Mathura Road, New Delhi 110 044
India

SAGE Publications Asia-Pacific Pte. Ltd.
18 Cross Street #10–10/11/12
China Square Central
Singapore 048423

President: Mike Soules
Associate Vice President and Editorial Director:
   Monica Eckman
Publisher: Erin Null
Content Development Editor: Jessica Vidal
Production Editor: Tori Mirsadjadi
Copy Editor: Liann Lech
Typesetter: Integra
Proofreader: Susan Schon
Indexer: Integra
Cover Designer: Scott Van Atta
Marketing Manager: Deena Meyer

Printed in the United States of America.

Library of Congress Cataloging-in-Publication Data

Names: Kobett, Beth McCord, author. | Fennell, Francis M., 1944- author. | Karp, Karen S., author. | Andrews, Delise, author. | Knighten, Latrenda, author. | Shih, Jeff, author.

Title: Classroom-ready rich math tasks, grades K-1 : engaging students in doing math / Beth McCord Kobett, Francis (Skip) Fennell, Karen S. Karp, Delise Andrews, Latrenda Knighten, and Jeff Shih.

Description: Thousand Oaks, California : Corwin, [2021] | Includes bibliographical references.

Identifiers: LCCN 2020057747 | ISBN 9781544399102 (paperback) | ISBN 9781544399119 (adobe pdf) | ISBN 9781071841235 (ebook) | ISBN 9781071841242 (ebook)

Subjects: LCSH: Mathematics--Study and teaching (Kindergarten) | Mathematics--Study and teaching (Elementary)

Classification: LCC QA135.5 K626 2021 | DDC 372.7/049--dc23

LC record available at https://lccn.loc.gov/2020057747

This book is printed on acid-free paper.

21 22 23 24 25 10 9 8 7 6 5 4 3 2

Contents

Visit the companion website at
resources.corwin.com/classroomreadymath/K-1
for downloadable resources.

Preface

We know that teachers work incredibly hard to make their mathematics lessons meaningful, challenging, and accessible to their students. We were inspired by a teacher we worked with when she told us:

> It seems like I spend a lot of time, often in the evenings, thinking about and then creating or locating, and usually revising, math tasks for my students. Having my students engage in challenging mathematics tasks is important to my planning and teaching—every day!

> (Elementary school teacher)

We also know how valuable time is and how many, if not most, teachers find it to be in very short supply. For some time, we have felt that teachers need to have access to mathematics tasks that fully represent the most important mathematical topics and standards they are responsible for teaching, and that will truly engage their students in doing the mathematics they are learning.

This book is designed to address these concerns! Here's our plan. We begin by unpacking a *doing-math task* (Chapter 1). The chapter defines and addresses the importance of *doing-math tasks* and why such rich, engaging, and high-level tasks are important. The chapter then considers issues related to planning for task implementation, with particular emphasis on task selection and teaching.

In Chapters 2 and 3 we address major planning and instructional issues relative to the use of rich mathematics tasks, the importance of which, to an extent, pushed us to write this book and the companion Grades 2–3 and Grades 4–5 volumes. Our point, our concern—it's not just the task! We have been in too many classrooms and had too many conversations with our teacher and teacher leader colleagues and friends about the need, the necessity, to truly recognize and value the importance of planning for and implementing such tasks. Planning considerations include ensuring that every math task addresses important mathematics standards and related mathematical practices. We also discuss the characteristics of effective mathematics teaching practices, including, for example, the use of multiple representations. Our approach is to address issues related to planning (Chapter 2) and instruction (Chapter 3) using the task template presented in Chapters 4–16. The template defines a *doing-math task*-based lesson, which, as noted, considers much more than just locating and presenting a mathematics task.

As Chapter 2's title indicates, the chapter lays the groundwork for teaching with rich *doing-math tasks*, with attention to planning-related essentials. These features include mathematics vocabulary, materials, grouping approaches, strategies for engaging your students in the task, and other anticipatory considerations related to preparing to implement a task-based lesson. We also note the importance of setting norms for task-based inquiry and provide specific planning and teaching-related suggestions regarding the importance of student and task access and equity; opportunities for productive struggle as your students engage in *doing-math tasks*; alternate learning environments, including online teaching; and strengths spotting—being able to recognize student strengths and considering them as the launch point for your work with rich *doing-math tasks*.

Chapter 3 is all about implementing a *doing-math task*-based lesson. This process begins with suggestions regarding the launching of a task-based lesson, which includes, but is not limited to, the following techniques: See, Think, and Wonder; Notice and Wonder; and the Three-Read Protocol. The facilitation of a *doing-math task* lesson includes classroom considerations related to grouping students for engaging in the task; classroom discourse, including the use of purposeful questioning; and monitoring the task lesson's progress using the following five classroom-based formative assessment techniques: Observations, Interviews, Show Me, Hinge Questions, and Exit Tasks. Our task lesson template considers lesson closure as an opportunity to make visible the mathematics students learn. The chapter suggests teaching strategies to involve students in sharing their work and learning from each other. Finally, the chapter discusses the importance of taking the time to reflect on the planning and implementation of each *doing-math task* lesson.

Chapters 4–16 are chock-full of 56 rich *doing-math tasks* that your students will want to engage in as they make sense of the mathematics they are learning. The tasks provided address important mathematics standards within the content strands appropriate for Grades K and 1, and are presented using the task-based lesson template discussed earlier and described within Chapters 2 and 3.

Finally, Chapter 17, titled Your Turn, is *your* opportunity to reflect on what you have learned and the tasks you have implemented, and to explore recommendations for selecting, adapting, and creating tasks in the future. The chapter also examines responses to a sampling of frequently asked questions that teachers have when teaching with mathematics tasks.

Well, let's move on! Get ready for some lively, cooperative, mathematical discussions that are sure to reveal your students' unique and powerful mathematical insights as you and they become immersed in *doing-math tasks*.

Acknowledgments

From Beth McCord Kobett: It is an impossible task to describe my gratitude to the people who have supported me and lifted me up to do things that I never thought I could do, but here it goes … . I am forever grateful to my husband, Tim, who, from the beginning, has patiently supported and championed my interests and passion for education. My deep appreciation to our daughters, Hannah and Jenna, for their enduring love and support. To an incredible author team, I express my gratitude to Skip and Karen for anchoring this project with their marvelous wisdom and expertise, and to Jeff, Latrenda, and Delise for their diligence and brilliance in making this work come alive. To my preservice teachers, graduates, and colleagues, I am in constant awe of your dedication, creativity, and passion for teaching and learning—thank you for inspiring me every day.

From Francis (Skip) Fennell: To Nita, Brett, Heather, and Stacey for all of the support and patience! And to all of my former students and colleagues at McDaniel College, thanks for regularly providing me with the support, feedback, and opportunity to, I hope, make a difference. I would also like to thank all co-authors, with particular thanks to Dr. Beth Kobett for her vision and diligence related to the importance of defining and implementing *doing-math task* lessons.

From Karen S. Karp: I wish to thank the co-authors of this book, from whom I have learned a great deal throughout the writing process. I especially would like to thank Beth Kobett for inviting me to participate and being our courageous leader.

From Delise Andrews: I wish to thank Dr. Beth Kobett for inspiring this work and giving me the opportunity to be a part of it; Ms. Sally Dunham, who taught me everything I know about teaching; Dr. Matt Larson and Dr. Jim Lewis, both pivotal mentors to me in mathematics education; and most of all my parents, Jerry and Kay Andrews, who faithfully trained me up in the way I should go.

From Latrenda Knighten: I wish to thank my family: My parents, Dianne and Randolph (deceased) Knighten, for their constant support and encouragement. Most importantly, thanks for knowing when to say, "If you've never tried it before, how do you know you can't do it or won't like it?" I also wish to thank my four sisters, who aren't educators, but they have earned their "teacher wings" by volunteering in my classroom, preparing materials for my students, and helping prepare materials for teacher workshops. I would also like to thank my extended family—the many educators, parents, and students who nurtured, supported, and guided me when needed and also served as mentors, collaborators, and cheerleaders.

From Jeff Shih: Thank you to Beth Kobett for guidance, leadership, and a steadying influence. Thank you to the Dream Author Team of Latrenda, Delise, Karen, and Skip: it was an honor to work with you all. And of course, thank you to Meg, Abby, and Penelope.

From the authors: We are very thankful for our publisher at Corwin, Erin Null. She championed this work many years ago when this project was just a nub of an idea. Erin's superhero ability to first navigate hefty, somewhat unformed ideas and target what is practical and thoughtful with a laser-like focus is truly a marvelous talent to behold. We also want to thank Jessica Vidal for her enduring commitment and devotion to transmitting this book from start to finish and making the work shine. To our project editor, Tori Mirsadjadi, for her expertise, patience, and top-knotch organizational skills. To the entire Corwin team, we thank you for your careful eyes and hands each step of the way.

We are grateful to students, teachers, and families everywhere who show up to learn and teach mathematics in an uncertain world. We hope that you will take a moment to celebrate the brilliance and strengths that each of you hold individually and collectively.

Publisher's Acknowledgments

Corwin gratefully acknowledges the contributions of the following reviewers:

Jonathan D. Bostic
Associate Professor of Mathematics Education
Bowling Green State University
Bowling Green, OH

Shelley Dickson
District Math Specialist, Primary
Fayette County Public Schools
Lexington, KY

Kristine Gettelman
Mathematics Specialist/Consultant
Mathematics Institute of Wisconsin
Waukesha, WI

Terina Legge
K–6 Mathematics Program Specialist
Newfoundland and Labrador English School District
Newfoundland and Labrador, Canada

Thomas Roberts
Assistant Professor
Bowling Green State University
Bowling Green, OH

About the Authors

Beth McCord Kobett is a professor in the School of Education at Stevenson University, where she works with preservice teachers and leads professional learning efforts in mathematics education both regionally and nationally. She currently serves on the board of directors for the National Council of Teachers of Mathematics and is the former president of the Association for Maryland Mathematics Teacher Educators. She is a former classroom teacher, elementary mathematics specialist, adjunct professor, and university supervisor. At the undergraduate level, Dr. Kobett teaches early childhood, elementary, and middle school mathematics methods and content courses. She has coauthored nine peer-reviewed teaching and coaching mathematics books in addition to numerous book chapters and articles. Dr. Kobett is a recipient of the Mathematics Educator of the Year Award from the Maryland Council of Teachers of Mathematics and the Johns Hopkins University Distinguished Alumni award. Dr. Kobett also received the Johns Hopkins Excellence in Teaching Award as a part-time instructor and Stevenson University's Excellence in Teaching Award as both an adjunct and full-time faculty member.

Francis (Skip) Fennell, PhD, DHL, is emeritus as the L. Stanley Bowlsbey Professor of Education and Graduate and Professional Studies at McDaniel College in Maryland, where he also directed the Elementary Mathematics Specialists and Teacher Leaders Project. He is a former classroom teacher, principal, and supervisor of instruction, and past president of the Association of Mathematics Teacher Educators (AMTE), the Research Council on Mathematics Learning (RCML), and the National Council of Teachers of Mathematics (NCTM). He is a recipient of the Mathematics Educator of the Year Award from the Maryland Council of Teachers of Mathematics (MCTM), the Glenn Gilbert National Leadership Award from the National Council of Supervisors of Mathematics (NCSM), the Excellence in Leadership and Service in Mathematics Teacher Education Award from the Association of Mathematics Teacher Educators (AMTE), the James W. Heddens Distinguished Service Award from the Research Council on Mathematics Learning (RCML), and the Lifetime Achievement Award from the Maryland Council of Teachers of Mathematics (MCTM) and the National Council of Teachers of Mathematics (NCTM). In 2018, he received an honorary Doctor of Humane Letters degree from McDaniel College.

Karen S. Karp is a professor in the School of Education at Johns Hopkins University. Previously, she was a professor of mathematics education at the University of Louisville. She is a former member of the board of directors of the National Council of Teachers of Mathematics and a former president of the Association of Mathematics Teacher Educators. Dr. Karp is a recent recipient of the NCTM Lifetime Achievement Award for Distinguished Service to Mathematics Education. She is a member of the author panel for the *What Works Clearinghouse Practice Guide on Assisting Students Struggling with Mathematics* for the U.S. Department of Education Institute of Educational Sciences. She is the author or coauthor of approximately 20 book chapters, 50 articles, and 40 books, including *Elementary and Middle School Mathematics: Teaching Developmentally, Developing Essential Understanding of Addition and Subtraction for Teaching Mathematics,* and *Inspiring Girls to Think Mathematically*. She holds teaching certifications in elementary education, secondary mathematics, and K–12 special education.

Delise Andrews is the 3–5 Mathematics Coordinator for Lincoln Public Schools in Lincoln, Nebraska. During her career, she has worked in both rural and urban districts and has taught mathematics to students at every age from kindergarten through the eighth grade, undergraduate mathematics methods, and graduate-level courses for teachers of mathematics. Delise is a recipient of the Presidential Award for Excellence in Mathematics and Science Teaching and a Robert Noyce Master Teaching Fellow. She is also an active member of NCTM, serving as a past member and chair of the Professional Development Services Committee, member of regional conference committees and chair of the St. Louis annual conference committee, and an NCTM Professional Services facilitator.

Latrenda Knighten is a District Elementary Mathematics Instructional Coach in the Office of Curriculum & Instruction and Professional Development for the East Baton Rouge Parish School System in Baton Rouge, Louisiana. She has been an educator for more than 30 years, during which she has been a classroom teacher, an elementary science specialist, and an elementary mathematics coach. Latrenda is an active member of many professional organizations where she has served in leadership roles for several local, state, and national organizations. This includes serving as past president of the Louisiana Association of Teachers of Mathematics (LATM) and the Baton Rouge Area Council of Teachers of Mathematics (BRACTM). She currently serves as the National Council of Supervisors of Mathematics Southern Region 2 Team Leader for Louisiana and Secretary for the Benjamin Banneker Association, Inc. Latrenda is also a past member of the National Council of Teachers of Mathematics Board of Directors.

Jeff Shih is a professor of mathematics education at the University of Nevada, Las Vegas, and currently serves on the Board of Directors for the National Council of Teachers of Mathematics (NCTM). He has served on the editorial boards for the *Journal for Research in Mathematics Education, Cognition and Instruction, Elementary School Journal, Mathematics Teacher Educator,* and *Investigations in Mathematics Learning.* He was the coordinator for the Nevada Collaborative Teaching Improvement Program for over 10 years and also led the Association of Mathematics Teacher Educators (AMTE) Service, Teaching, & Research (STaR) Fellows Program.

Doing-Math Tasks

What Are They, Why Are They Important, and How Do I Plan for Implementation?

In this chapter, you will explore the idea of *doing-math tasks* and their importance in an instructional model that includes engaging, high-cognitive-demand tasks as part of your overall mathematics teaching. By the end of this chapter, you will

- understand the need for and importance of *doing-math tasks*,

- explore how *doing-math tasks* increase your students' mathematical understandings, and

- explore the research behind selecting and implementing mathematics tasks.

As students learn mathematics, they must be actively involved in learning experiences that both challenge and engage them. The work students do—which is, in practice, defined by the *tasks* teachers assign—determines how they think about solving problems. This level of engagement is how they come to understand concepts and ideas, how they give meaning to mathematics. But tasks are essentially how we would want to picture students actually "doing math," right? We want to see students truly engaged in activities—every day—that help them to connect with concepts and skills and deepen their understanding of the mathematics they are learning. One way to do this is by providing experiences that develop the learner's ability to apply their mathematical knowledge in novel problem-based settings, a life skill expected of 21st century learners (National Research Council, 2012). Such learning opportunities are best supported with what we, in this book, describe as *doing-math tasks*. Stated briefly, *doing-math tasks*

> " The ways students think about solving problems are governed by the tasks we assign. "

> **Doing-math tasks** require your students to explore, yet also to self-monitor, their thinking as they mentally retrieve prior instructional experiences while working through a task.

require your students to explore, yet also to self-monitor, their thinking as they mentally retrieve prior instructional experiences while working through a task. And, yes, such tasks should challenge and involve your students in relevant contexts (National Council of Teachers of Mathematics, 1991; Smith & Stein, 1998). The classroom environment where students are *doing-math* by engaging in meaningful tasks is anchored in the critical role of teachers in creating, locating, and implementing them. We'll explore the elements of such tasks in more depth and demonstrate how this book supports your role in becoming a connoisseur and developer of high-quality tasks yourself. Then we will share an amazing collection of tasks that will absorb your students in regularly *doing-mathematics*!

Here's the challenge. *Doing-math tasks* are typically not what you may find in your school's prescribed set of mathematical instructional materials, which is why this book's focus on *doing-math tasks* is so important. *Doing-math tasks* require student exploration. They help students develop and implement solution strategies that draw on their prior knowledge and learning experiences and translate them into use for new situations (Smith & Stein, 1998). That said, let's consider the reality of the classroom. A recent RAND Corporation study of 2,873 U.S. teachers found that almost 100 percent reported that they used instructional materials "I developed and/or selected myself" (Opfer et al., 2017). Such curation of curriculum from a wide array of resources can actually result in entirely different lessons within the same grade, certainly a potential for curricular confusion, as well as a possibly mismatched set of mathematical progressions across an entire school. In addition, the online or print gathering of a self-selected collection of lessons is certainly not a reliable path to a clear, cohesive, or equitable instructional plan. This concern is addressed directly within NCTM's (2020a) *Catalyzing Change in Early Childhood and Elementary Mathematics*:

> The danger of teachers creating their own daily lessons with instructional resources found at random through search engines is that mathematics topics are treated as isolated containers of ideas to master in a lesson or to experience through a "fun" activity. The likely result is instruction that is not deep, coherent, or aligned with a carefully crafted developmental learning sequence. The progressive nature of mathematics learning demands coherent instructional experiences that build and connect to one another, which is best accomplished through high-quality mathematics instructional materials. (p. 39)

Like you, we recognize the importance of students being engaged in doing the mathematics they are learning. We also know that locating, developing or adapting, planning for, and implementing such *doing-math tasks* can be a daunting and time-consuming task on top of all the other subject areas you are planning for. If you are one of the hundreds of thousands of elementary classroom teachers who spend hours searching online for truly engaging mathematics activities, or if you want to differentiate more to meet the needs of your students, or if you fret over the fact that you simply *don't have the time* to do all of "this," this book's for you! The mathematical tasks presented in Chapters 4–16 are intended to be integral elements of mathematics lessons at the grade levels you teach. These task-based lessons are neither intended to replace your major curriculum resources—textbook or otherwise—nor to be used every day; rather, they are intended to be "distinctive tasks" that will engage your students as they explore, understand, analyze, apply, and, yes—more than occasionally—productively struggle while they truly are *doing-mathematics*.

> In this book, you will find 56 kindergarten and first-grade tasks with accompanying resources.

What Is a *Doing-Math Task*?

Long-term research about the use of well-designed mathematical tasks as an important component of mathematics classroom or school-wide instruction led to the development of an influential guide for evaluating the quality of such tasks (Figure 1.1). This guide, designed by Smith and Stein (1998), outlines characteristics of effective tasks on a continuum from low-level cognitive demand to high-level cognitive demand.

Figure 1.1 Mathematics Task Analysis Guide

Level of Demands

Lower-level demands (memorization):

- Involve either reproducing previously learned facts, rules, formulas, or definitions or committing facts, rules, formulas, or definitions to memory

- Cannot be solved using procedures because a procedure does not exist or because the time frame in which the task is being completed is too short to use a procedure

- Are not ambiguous. Such tasks involve the exact reproduction of previously seen material, and what is to be reproduced is clearly and directly stated.

- Have no connection to the concept or meaning that underlies the facts, rules, formulas, or definitions being learned or reproduced

Lower-level demands (procedures without connections):

- Are algorithmic. Use of the procedure either is specifically called for or is evident from prior instruction, experience, or placement of the task.

- Require limited cognitive demand for successful completion. Little ambiguity about what needs to be done and how to do it.

- Have no connection to the concepts or meanings that underlie the procedure being used

- Are focused on producing correct answers instead of on developing mathematical understanding

- Require no explanations or explanations that focus solely on describing the procedure used

Higher-level demands (procedures with connections):

- Focus students' attention on the use of procedures for the purpose of developing deeper levels of understanding of mathematical concepts and ideas

- Suggest explicitly or implicitly pathways to follow that are broad general procedures that have close connections to underlying conceptual ideas as opposed to narrow algorithms that are opaque with respect to underlying concepts

- Usually are represented in multiple ways, such as visual diagrams, manipulatives, symbols, and problem situation. Making connections among multiple representations helps develop meaning.

- Require some degree of cognitive effort. Although general procedures may be followed, they cannot be followed mindlessly. Students need to engage with conceptual ideas that underlie the procedures to complete the task successfully and that develop understanding.

Higher-level demands (doing mathematics):

- Require complex and nonalgorithmic thinking—a predictable, well-rehearsed approach or pathway is not explicitly suggested by the task, task instructions, or a worked-out example

- Require students to explore and understand the nature of mathematical concepts, processes, or relationships

- Demand self-monitoring or self-regulation of one's own cognitive processes

- Require students to access relevant knowledge and experiences and make appropriate use of them in working through the task

- Require students to analyze the task and actively examine task constraints that may limit possible solution strategies and solutions

- Require considerable cognitive effort and may involve some level of anxiety for the student because of the unpredictable nature of the solution process required

Source: These characteristics are derived from the work of Doyle on academic tasks (1988) and Resnick on high-level-thinking skills (1987), the *Professional Standards for Teaching Mathematics* (NCTM, 1991), and on the examination and categorization of hundreds of tasks used in QUASAR classrooms (Stein, Grover, and Henningsen 1996; Stein, Lane, and Silver, 1996).

Here are examples of a task in each level of demand category at the kindergarten and first-grade levels (Fennell et al., 2017):

Lower-Level Demands—Memorization

- *When the student is shown a collection of shapes, ask "Which shape is a rectangle?"*

Lower-Level Demands—Procedures Without Connections

- *3 + 5 = ?*

Higher-Level Demands—Procedures With Connections

- *Caleb had 8 acorns and found 4 more acorns to add to his collection. Jamillia had 7 acorns and found 6 more acorns to add to her collection. Prove who has more acorns using cubes or drawings.*

Higher-Level Demands—Doing-Mathematics

- *Create a word problem for the following: 12 + 17 = ? Solve the problem. Then be ready to share how you solved it.*

> "Your conscious decision making about the tasks you select and use in the mathematics classroom makes a demonstrable difference in student learning!"

Your conscious decision making about the tasks you select and use in the mathematics classroom makes a demonstrable difference in student learning! Your choice of high-quality tasks and implementation using high-quality instructional strategies results in high student learning gains (Stein & Lane, 1996), as seen in Figure 1.2.

Figure 1.2 Eventual Learning Results of Differing Levels of Task Quality and Implementation

Task Quality	Implementation	Results
Low	High or low	Low
High	Low	Moderate
High	High	High

Source: Kobett, B. & Karp, K. (2020). *Strengths-based Teaching and Learning in Mathematics: 5 Teaching Turnarounds for Grades K-6.* Newbury Park, CA: Corwin and Reston, VA: NCTM.

When you expect to regularly implement high-level tasks, the quality of the task is important. Similarly, the actual classroom implementation of a high-quality, *doing-math task* is of related consequence. For example, if the task is of low quality, the results will be low, regardless of how masterful your instruction is. By low-quality tasks, we mean reduced cognitive demand that is potentially watered down, focused on memory or procedures, and generally lacking connections to important ideas. Low implementation means there is little questioning, a lack of attention to students sharing their thinking, and a surface level of attention to developing the mathematical meaning. Similarly, you can have a high-quality task combined with low implementation, which also doesn't produce what you want. The highest-level results come from the combination of a high-level task and high-level implementation. So, what do teachers who are using high-level tasks with high levels of implementation "look and act like"? These teachers

- ask questions that have an elevated cognitive demand,

- promote both student-to-teacher discourse as well as student-to-student discussion,

- arrange students in groups to compare solution strategies, and

- make the mathematics evident and visible by using their students' own mathematics thinking to advance their students' understanding.

But let's be cautious. Stein et al. (1996) suggest that sometimes with the best intentions, we think the ideal approach to implementing a challenging task might be to reduce the cognitive demand. This approach can sometimes inadvertently result in overexplaining, focusing on procedures without emphasizing their meaning, jumping in to demonstrate or model for students before they have opportunities to fully explore the problem, or dissecting a task into such small bits that the larger task vanishes. Situating students as "thinkers" as well as "doers" during lesson implementation creates an environment that promotes success and likewise positions students' strengths as the forefront of task design. That's precisely what the tasks in this book are positioned to do.

> " Situating students as "thinkers" as well as "doers" during lesson implementation creates an environment that promotes success and likewise positions students' strengths at the forefront of task design. "

WHY IS THE SELECTION OF HIGHER-LEVEL *DOING-MATH TASKS* IMPORTANT?

The first step in this process is the selection of these higher-level tasks, so let's look at why that is of great consequence. Figure 1.3 breaks down the research on how effective use of *doing-math tasks* is both a curricular and an instructional essential. Then it describes what implications we need to consider as we plan for and implement them, including specific examples from the tasks in this book.

Figure 1.3 From Research to Practical Examples

Research	Classroom Implication	Task Examples
Student learning of mathematics is greatest in classroom settings where the tasks encourage high-level student thinking and reasoning, and the learning is least in classrooms where the tasks are typically procedural (Boaler & Staples, 2008; Hiebert & Wearne, 1993; Stein & Lane, 1996).	This finding validates the importance of selecting mathematical tasks that challenge your students. The tasks we provide in this book are deliberately designed to engage your students in *doing-mathematics* at a high level of challenge.	In Task 53, **What's My Attribute?** (p. 266) students explore attributes that have been sorted and determine the sorting rule. *Sheena sorted the shapes below into two groups. What attributes did she use to sort the shapes? Label each group. Draw another shape that belongs in one of the groups. Why does it belong in that group?*

(Continued)

Figure 1.3 (Continued)

Research	Classroom Implication	Task Examples
Tasks should vary in opportunities for student thinking and learning (Hiebert et al., 1996; Stein et al. 2009).	Each task you locate, create, or adapt is perceived differently by the intended audience—your students. So, when considering access and equity for each and every child, you want to provide relevant and sometimes modified tasks tailored specifically to student needs without reducing the rich features of the task. Modifications could include changing contexts to reflect students' experiences, using sentence stems to support language access, and providing choice of solution pathways.	In Task 21, **Splat and Split** (p. 127), students are organized in pairs to play a game to find combinations for 10 by dropping counters onto a game board. Collaborative tasks like this one encourage multilingual learners to discuss the task and represent their understanding using concrete representations that they can record. *Madeline and Hector found an old box lid in the basement. They decided to create a game. They drew a line across the middle of the lid and then took turns dropping handfuls of 10 counters over the lid. Madeline got points for every counter that landed on one side of the lid. Hector got points for every counter that landed on the other side of the lid. If they drop 10 counters each time, how many points might each person get?*
Mathematical tasks should promote and challenge thinking. This expectation often makes higher-level cognitive-demand tasks more intensive to plan for and implement (Smith & Stein, 2018; Stein et al., 1996; Stigler & Hiebert, 2004).	High-demand tasks may take more time to implement. Because these tasks often involve using procedures with connections, and they may include use of new instructional tools and different mathematical representations, plan for them to take the majority of your daily mathematics instructional time.	In Task 37, **Make the Bear Family** (p. 189), students work collaboratively to construct the lengths of the bear family using the length of Goldilocks as the unit. This task is rich in language and prompts students to test their ideas. *Baby Bear is twice as tall as Goldilocks. Mama Bear is four times as tall as Goldilocks. Papa Bear is six times as tall as Goldilocks. Use the Goldilocks picture to find the heights of Baby, Mama, and Papa Bear. Then make each member of the Bear family. How tall is each member of the Bear family? Be prepared to show how you know.* *Use the Goldilocks unit or the Baby Bear unit to make a tree for the forest where the bears live. Your tree has to be at least two Goldilocks units tall. How many Goldilocks units tall is your tree? How many Baby Bear units tall is your tree? Be prepared to show how you know how tall your tree is.*

Reflect

1. As you read the research, classroom implications, and examples, what resonated with you? Why?

2. When you think about the phrase *doing-mathematics* as specifically applied to kindergarten and first-grade students, what does it mean to you?

3. How do you address student access to mathematics learning? Ask yourself:

   • How can this task be accessed by each of my students?

   • Will this task fully engage each of my students in *doing-math*? Consider the students' strengths and needs.

   • What are the task's entry points for my students?

4. What are issues you face as you plan for and implement mathematics tasks with a high cognitive demand?

How Do I Plan to Implement a *Doing-Math Task*?

As you recognize the importance of *doing-math tasks* that truly engage your students, the question of "How do I do this more frequently?" logically emerges. One teacher told us, "I understand the importance of mathematics tasks, but wonder how I can literally 'find' such tasks on a regular basis." Fortunately, we have written an entire book of tasks that addresses important mathematics topics in kindergarten and first grade. But if you still find yourself tempted to hunt for more, there are several steps to finding and selecting—or adapting—and then implementing these kinds of tasks. The first two steps go hand-in-hand and must happen concurrently.

SELECTING THE TASK

First, let's consider the following decision points around task selection. These choices include answering the questions in Figure 1.4.

Figure 1.4 Making Decisions About Task Selection

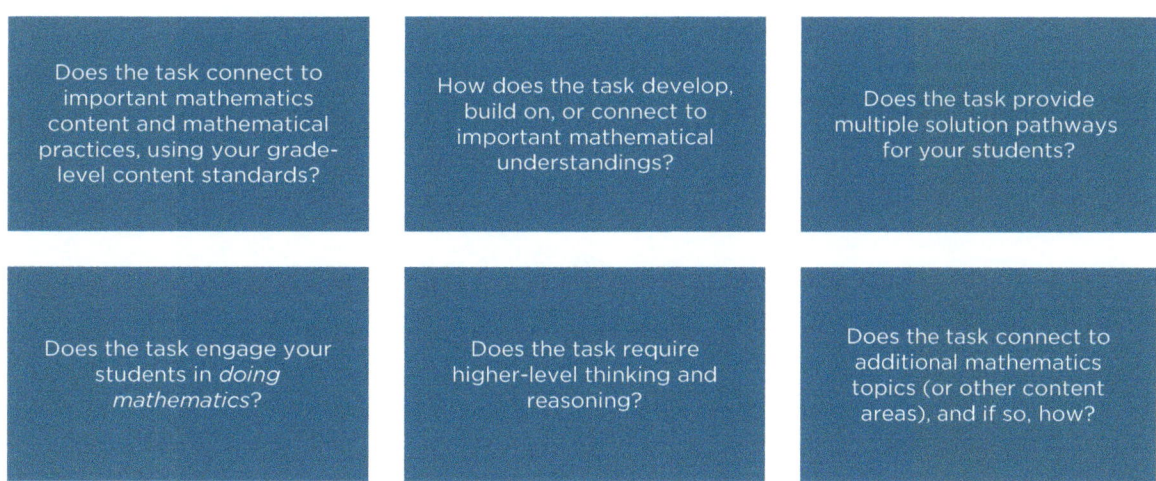

Does the task connect to important mathematics content and mathematical practices, using your grade-level content standards?

How does the task develop, build on, or connect to important mathematical understandings?

Does the task provide multiple solution pathways for your students?

Does the task engage your students in *doing mathematics*?

Does the task require higher-level thinking and reasoning?

Does the task connect to additional mathematics topics (or other content areas), and if so, how?

Next, consider how you would answer the following questions as you prepare to implement the task in your classroom (Figure 1.5).

Figure 1.5 Task Implementation

How will you position the task within a lesson? How much time is needed for your students to engage with the task during the lesson's launch, for your facilitation of the task experience, and for the lesson's close?

How and when will students use different representations (e.g., physical, pictorial, symbolic) as they engage in the task?

How and when will you make the mathematics visible by explicitly connecting student representations and other examples of student work to the task?

How will students present their solution strategies and the task's actual solution(s)?

How and when will you build in feedback to students regarding their performance on the task?

Summing Up

Mathematical tasks that engage students in actually *doing-math* provide reasoning and problem-solving opportunities that are integral to the development of student thinking and understanding. Such high-level or *doing-math tasks* must become everyday opportunities for all students.

Where are we in the process and what's next? This chapter began by noting the importance of mathematical tasks; considered the levels of demand of mathematics tasks and research findings on the need for their use; and, in particular, highlighted this book's specific area of focus: *doing-math tasks.* You reviewed important considerations related to the use of math tasks instructionally. Finally, you examined classroom implications and challenges related to your selection and implementation of *doing-math tasks.* This chapter's focus on defining the importance of *doing-math tasks* and presenting considerations for their use naturally leads to the intent of the next chapter (Chapter 2), which focuses on a careful consideration of the elements of our task-based instructional model, and Chapter 3, which zeroes in on the implementation of the *doing-math tasks* we have provided.

In the tasks we have created for this book, we have designed structures that address these important selection and implementation questions, but you may still need to ask yourself how these structures and questions fit into your classroom environment. Chapters 2 and 3 consider task selection and implementation in more depth. Chapter 17 explores options for adjusting or adapting the tasks to fit the specific needs of your students.

Professional Learning/Discussion Questions

Read and discuss the following questions with your grade-level teaching team or with teams across multiple grade levels.

- In your own words, describe the benefits of engaging your students in high-cognitive-demand mathematics tasks.

- What is most important to consider as you select/create and implement a mathematics task in your classroom?

- How do you find the mathematics tasks that you currently use in your classroom?

- What is a *doing-math task*?

- What concerns you about locating, creating, or adapting mathematics tasks?

- What is the importance of your whole grade level or school using *doing-math tasks* as a team effort?

CHAPTER

2

Laying the Groundwork for Teaching With *Doing-Math Tasks*

In the first chapter, we explored what a mathematics task is and what it means when students are thoughtfully and meaningfully engaged in mathematics learning through the use of targeted tasks that align with mathematics standards. In this chapter, we will walk further down that path through some of the foundational elements that need to be in place in order for *doing-math task*-based instruction to meet its maximum potential. We will introduce you to and familiarize you with the components of the task template we use throughout the book to present the *doing-math tasks*. For this chapter, we explore all of the components that you will need to know to prepare to teach a *doing-math task*. Later, in Chapter 3, you will learn all about the components that have to do with implementing a *doing-math task*. In this chapter, you will

- explore mathematics content through grade-level standards and the mathematical practices and processes;

- understand the important role of mathematics vocabulary for students to know or develop;

- explore how *doing-math task*-based lessons and materials support students in representing their mathematical thinking;

- unpack task preparation by investigating the importance of strategically grouping students for *doing-math tasks* and establishing group and classroom norms; and

- determine how attention to access and equity, productive struggle, alternate learning environments, and student strengths spotting is addressed in the mathematics tasks.

Once you move into Chapters 4–16 of this book, you will find 56 *doing-math tasks* reflecting the important mathematics topics and standards that are central for Grades K and 1.

Effective Mathematics Teaching Practices

We can't consider the selection of mathematics tasks or begin to think through their implementation without first looking at the importance of specific teaching practices that have the potential to make a difference in students' learning and performance. This emphasis is needed because student learning of mathematics "depends fundamentally on what happens inside the classroom as teachers and learners interact over the curriculum" (Ball & Forzani, 2011a p.17). These "high-leverage" practices (Ball & Forzani, 2011b), reconceptualized in NCTM's *Principles to Actions* (2014b) as Effective Teaching Practices (Figure 2.1), are research-informed practices that promote students' mathematical learning. The NCTM Effective Teaching Practices guided the development of the tasks and the task-based lesson format used in this book.

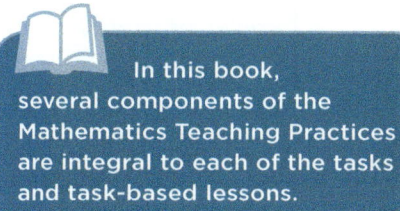

In this book, several components of the Mathematics Teaching Practices are integral to each of the tasks and task-based lessons.

Figure 2.1 NCTM's Effective Mathematics Teaching Practices

Effective Mathematics Teaching Practices	
Establish mathematics goals to focus learning.	Effective teaching of mathematics establishes clear goals for the mathematics that students are learning, situates goals within learning progressions, and uses the goals to guide instructional decisions.
Implement tasks that promote reasoning and problem solving.	Effective teaching of mathematics engages students in solving and discussing tasks that promote mathematical reasoning and problem solving and allow multiple entry points and varied solution strategies.
Use and connect mathematical representations.	Effective teaching of mathematics engages students in making connections among mathematical representations to deepen understanding of mathematics concepts and procedures and as tools for problem solving.
Facilitate meaningful mathematical discourse.	Effective teaching of mathematics facilitates discourse among students to build shared understanding of mathematical ideas by analyzing and comparing student approaches and arguments.
Pose purposeful questions.	Effective teaching of mathematics uses purposeful questions to assess and advance students' reasoning and sense making about important mathematical ideas and relationships.
Build procedural fluency from conceptual understanding.	Effective teaching of mathematics builds fluency with procedures on a foundation of conceptual understanding so that students, over time, become skillful in using procedures flexibly as they solve contextual and mathematical problems.
Support productive struggle in learning mathematics.	Effective teaching of mathematics consistently provides students, individually and collectively, with opportunities and supports to engage in productive struggle as they grapple with mathematical ideas and relationships.
Elicit and use evidence of student thinking.	Effective teaching of mathematics uses evidence of student thinking to assess progress toward mathematical understanding and to adjust instruction continually in ways that support and extend learning.

Source: National Council of Teachers of Mathematics. (2014b). *Principles to Actions: Ensuring Mathematical Success for All.* Reston, VA: Author.

To learn more about the Mathematics Teaching Practices and Standards for Mathematics Practices, check out the following resources:

 CHECK THESE OUT

Principles to Actions: Ensuring Mathematical Success for All
NCTM (2014b)

Principles to Actions: Professional Learning Toolkit
www.nctm.org/PtAToolkit/
NCTM (2017b)

Taking Action: Implementing Effective Mathematics Teaching Practices in K–Grade 5
NCTM (2017c)

Routines for Reasoning: Fostering the Mathematical Practices in All Students
Kelemanik and Lucenta (2016)

Putting the Practices Into Action: Implementing the Common Core Standards for Mathematical Practice, K–8
O'Connell and SanGiovanni (2013)

Beyond Answers: Exploring Mathematical Practices With Young Children
Flynn (2017)

Now that we have explored the critical elements for developing powerful mathematics teaching and learning experiences, let's take a look at how the *doing-math tasks* template is designed for you to begin task implementation right away. We will explore the first half of the *doing-math task* template (see Figure 2.2), which contains the following elements:

- Mathematics Standard(s)

- Mathematical Practice(s)

- Task

- Vocabulary

- Materials

- Task Preparation

The second half of the *doing-math task* template focuses on implementing and reflecting on the task, which we explore in depth in Chapter 3.

Figure 2.2 Components of the Doing-Math Task

Grade Level

Task Title

Task Topic

Mathematics Standard(s):
Mathematical Practice(s):
Task

Vocabulary	Materials

Task Preparation:
Launch:
Facilitate:
Close: Make the Math Visible
Post-Task Notes: Reflection & Next Steps

 This online resource can be found in Appendix A and is available for download at **resources.corwin.com/ClassroomReadyMath/K-1**

What Is the Mathematics Standard?

The mathematics focus we highlight here incorporates the mathematics content standard(s) and practices or processes that define students' expectations at each grade level. Whether your state relies on the Common Core State Standards for Mathematics (CCSS-M), an adapted version, or mathematics standards unique to your state, the large majority of states use a collection of standards grounded in research-based learning progressions that highlight what is known about how students develop mathematical knowledge, skills, and understandings. Regardless of each state's mathematics standards, the use of mathematics content and practice or process standards to drive curriculum and instruction represents a renewed effort to focus on strategies for teaching that cross specific topics or big ideas. This emphasis within each grade level allows teachers to cohesively deepen the way they use their time and energy in the classroom.

" **Teachers are central to students' success.** "

Teachers are central to students' success because "teachers need to understand the big ideas of mathematics and be able to represent important mathematical topics as coherent and connected" (NCTM, 2000, p. 17).

The dual use of our language around the Standards for Mathematical Practice (SMPs) and the mathematical processes allows us to be inclusive of many states who have made different decisions along the way. Here we focus on how mathematical practices and processes align so you can make knowledge-based decisions about the use of the tasks in this book. While the mathematics content standards focus on the important mathematics that students should learn, the National Council of Teachers of Mathematics (NCTM) Process Standards were developed first and they draw attention to the idea that "what we teach [in mathematics] is as important as how we teach it" (NCTM, 1991, p. 23). The Process Standards (NCTM, 2000) are as follows:

1. **Problem Solving:** Students use a repertoire of skills and strategies for solving a variety of problems and situations.

2. **Communication:** Students use mathematical language including terminology and symbols to express ideas precisely.

3. **Reasoning and Proof:** Students apply inductive and deductive reasoning skills to make, test, and evaluate statements to justify steps in mathematical procedures.

4. **Connections:** Students relate concepts and procedures from different topics in mathematics to one another and make connections between topics in mathematics and other disciplines.

5. **Representation:** Students use a variety of representations, including graphical, numerical, algebraic, verbal, and physical, to represent, describe, and generalize.

The SMPs (National Governors Association & Council of Chief State School Officers, 2010) were heavily influenced by both the NCTM Process Standards and the strands of mathematical proficiency specified in *Adding It Up* (National Research Council, 2001). The following SMPs highlight the thinking processes, dispositions, or habits of mind that students should exhibit while engaging in mathematics learning experiences:

1. **Make sense of problems and persevere in solving them.** Students work to understand the information given in a problem and the question that is asked. They use a strategy to find a solution and check to make sure their answer makes sense.

2. **Reason abstractly and quantitatively.** Students make sense of quantities and their relationships in problem situations.

3. **Construct viable arguments and critique the reasoning of others.** Students explain their thinking, justify, and communicate their conclusions both orally and in writing.

4. **Model with mathematics.** Students use and apply multiple representations, models, and symbols to make sense of the mathematics.

5. **Use appropriate tools strategically.** Students use a variety of concrete materials and tools to represent their thinking when solving problems.

6. **Attend to precision.** Students learn to communicate precisely with each other and explain their thinking using appropriate mathematical vocabulary.

7. **Look for and make use of structure.** Students look for and discover patterns and structure in their mathematics work.

8. **Look for and express regularity in repeated reasoning.** Students notice if calculations are repeated. Students use patterns to make generalizations.

> In this book, the mathematics content standards and SMPs (we know that those of you who use the Process Standards will be able to make the connections) have been identified for each mathematics task that is presented. However, you may want to consider and then select different SMPs that you feel need to be emphasized depending on the needs of your students. As you teach a task, you may want to ensure that the students are exhibiting these SMPs as they engage with the task.

The SMPs and the Process Standards are particularly important in task-based teaching, because the expectation is that you want to teach mathematics by having students analyze relationships, communicate mathematical ideas by engaging in discourse, develop multiple solution pathways, articulate and justify their mathematical reasoning, and apply mathematics to real-world situations.

Vocabulary

The role of vocabulary in a mathematics lesson is unique to the mathematics classroom.

Mathematics vocabulary can be highly specialized, and students may encounter particular vocabulary only in a mathematics classroom because it is not likely to occur in everyday conversation. For example, think about the last time the word *radius* came up in casual conversation! Student success hinges on knowing and using academic vocabulary, so having only surface-level knowledge of vocabulary does not promote conceptual competence (Thompson & Rubenstein, 2000).

> **"** Student success hinges on knowing and using academic vocabulary. **"**

Instruction related to mathematics vocabulary requires special attention that is different from the way vocabulary is taught in a reading lesson, which often focuses on direct instruction of word meanings before students read a passage or book. In mathematics, however, a focus on presenting the word first may lead to partial or even incorrect understandings. For example, Celena decided to pre-teach the definition of a triangle by showing a picture of a triangle (Figure 2.3) and stating, "A triangle is a shape with three sides and three angles."

Figure 2.3 Triangle

Later in the lesson, when students were asked to sort triangles, many did not identify the shapes in Figure 2.4 as triangles because they "looked different" from the original example provided.

Figure 2.4 Set of Three Triangles

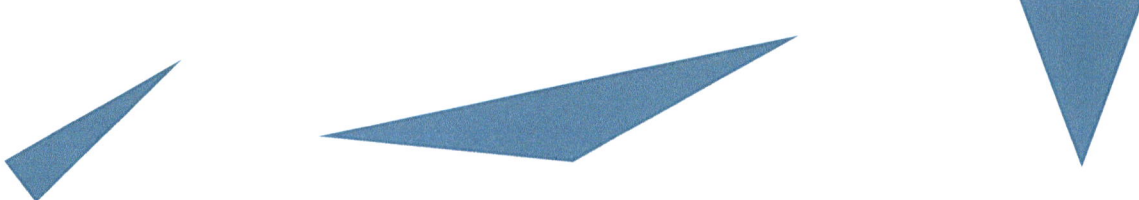

The next day, Celena decided to ask students to sort the triangles first and then develop a definition for a triangle using the students' experiences. The students constructed a class definition (Figure 2.5) that encompassed a variety of significant ideas.

Figure 2.5 Students' Ideas for a Triangle Definition

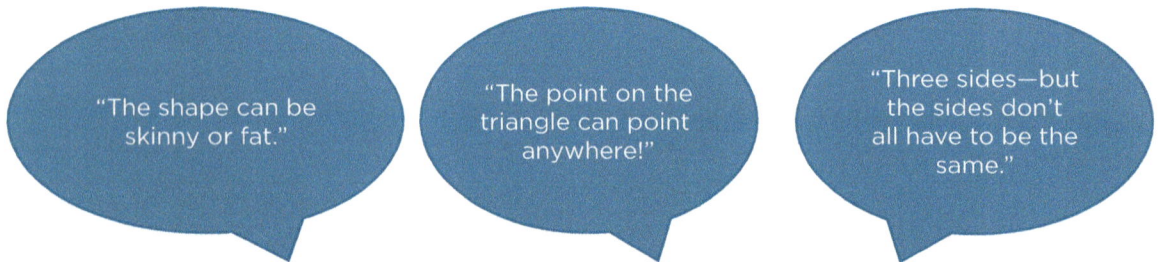

"The shape can be skinny or fat."

"The point on the triangle can point anywhere!"

"Three sides—but the sides don't all have to be the same."

While the student definitions are not standard textbook definitions, they do reflect their emerging understanding of the characteristics of the triangle and demonstrate how students exhibit Standard for Mathematical Practice (SMP) 6: Attend to precision: Mathematically proficient students try to communicate precisely to others. They try to use clear definitions in discussion with others and in their own reasoning (National Governors Association & Council of Chief State School Officers, 2010).

> In this book, important mathematics vocabulary is highlighted for all tasks. However, you know your students and your curriculum best and may want to loop in additional vocabulary. You can support students' building of mathematical vocabulary by asking them to explain their ideas throughout the lesson and, particularly, in the task lesson's close as you are making the mathematics learning visible to the students. As your students develop new understandings within a task-based lesson, add new words and corresponding visual representations, student definitions, examples, and non-examples to an interactive mathematics word wall.

To learn more about developing mathematics vocabulary, check out the following resources:

 CHECK THESE OUT

"Supporting Math Vocabulary Acquisition," *Teaching Children Mathematics*
Bay-Williams and Livers (2009)

The Math Pact, Elementary: Achieving Instructional Cohesion Within and Across Grades
Karp et al. (2020)

"Vocabulary Support: Constructing (Not Obstructing) Meaning," *Mathematics Teaching in the Middle School*
Livers and Bay-Williams (2014)

"'Weigh' to Go! Exploring Mathematical Language," *Mathematics Teaching in the Middle School*
Adams et al. (2005)

Materials

Materials such as manipulatives, chart paper, technology, or other means to represent ideas that are central to a *doing-math task*-based lesson support students in developing their understanding throughout the instructional experience. These materials are key as they provide students with the tools to create a variety of ways to represent their thinking. When students can see the connections between multiple representations, it is a signal that their conceptual understanding is in place.

CHOOSING AND USING MULTIPLE REPRESENTATIONS

When choosing representations, consider the types of manipulatives (physical or digital), drawings, equations, tools, and other resources that you may need to support student learning. The primary reason for using multiple representations in the classroom is to help students understand and access abstract mathematics concepts through three fluid stages: concrete, semi-concrete, and abstract (CSA) (Dougherty et al., 2016; Heddens, 1986; Laski et al., 2015; Van de Walle et al., 2019). Representations such as manipulative materials, for example, help students understand new mathematics concepts because "they serve as analogies; the things manipulated are symbols for the new, to-be-understood idea" (Willingham, 2017). At first, this may seem contradictory to the idea that manipulatives are concrete, and therefore learners will be able to immediately understand abstract ideas through the use of manipulatives. But as Ball (1992) stated in her classic article "Magical Hopes," there is an issue in believing that students immediately develop the ideas we hope they will from the use of manipulatives alone. The act of connecting concrete, semi-concrete, and abstract representations *together* is what can help students meaningfully associate a physical model to a visual representation of the mathematics situation *and* to the corresponding equation. The explicit linkages between these different representations in the CSA approach help develop students' mathematical ideas, and when students can articulate the relationships, it helps teachers know if the concept or procedure is understood.

> **When students can articulate the CSA relationship, it reveals whether they understand the concept or procedure.**

Teachers need to strategically plan for movement between the representations (Fyfe et al., 2014), to explain that mathematics instruction on a particular topic often begins with concrete manipulatives and then gently moves to a greater focus on just the abstract. However, we need to remember that we are asking students to use representations to represent ideas and integrate contexts within visuals, expressions, equations, and word problems. Teachers must strike a balance between ensuring that

In the tasks in this book, students are invited to use many different representations aligned with the CSA approach to represent mathematical ideas (e.g., regional fraction models, number lines, sketches, tables with numbers). As you plan for and prepare to teach the tasks, consider your students' needs and the suggested representations as you strategically select the most meaningful tools to develop important mathematical understandings throughout the task lesson.

their students understand the purpose of the representation for the task and not being too rigid in how students use the representation (Carbonneau et al., 2013). So, teachers might say, "Our task is about a quantity of markers. Consider using the base ten materials or cubes or drawings to represent the number of markers in the task."

To learn more about representations, check out the following resources:

CHECK THESE OUT

"Representation: An Important Process for Teaching and Learning Mathematics," *Teaching Children Mathematics*

Fennell and Rowan (2001)

"Star Students Make Connections: Discover Strategies to Engage Young Math Students in Competently Using Multiple Representations," *Teaching Children Mathematics*

Marshall et al. (2010)

"Facilitating Mathematical Practices Through Visual Representations," *Teaching Children Mathematics*

Murata and Stewart (2017)

Clothesline Math

clotheslinemath.com/

Shor (2017)

Teaching Student-Centered Mathematics: Developmentally Appropriate Instruction for Grades 3–5
Van de Walle et al. (2017)

The Mathematics Lesson-Planning Handbook, Grades K–2: Your Blueprint for Building Cohesive Lessons
Kobett et al. (2018)

Task Preparation

All worthwhile lessons based on a task require some level of preparation! As a part of the hundreds of decisions you make each day, you decide on the materials and representations you will use to support your students as they develop mathematical understandings. You also decide how you will group your students to best highlight their learning strengths and needs and establish norms for how students will work together to use strategies and develop solution pathways.

GROUPING YOUR STUDENTS TO ENGAGE IN THE TASK

In the words of one of our preservice teachers, Em, "Grouping students without data is just crazy!" We won't waffle, we strongly recommend that you arrange students strategically in heterogeneous learning groups or groups based on mixed strengths to engage in the book's tasks while you circulate to both observe and facilitate students' mathematical thinking and reasoning. We call this kind of instruction mixed-strength group instruction, meaning that the students are situated in a variety of small groups within the whole group and are instructed to complete the same or similar differentiated tasks while the teacher circulates the room, observes and facilitates thinking, and advances learning through probing questions (Kobett & Karp, 2020). During this time, students will have an opportunity to engage with partners or small groups throughout the lesson. Back to Em's point, however, teachers need to use data (see the section on Formative Assessment in Chapter 3 on p. 37) about what students already know about this topic and the solution pathways that they tend to use to assign students strategically to work in pairs or small groups so that their individual strengths shine and, at the same time, support one another to reason about the mathematics they are learning.

We are aware that many school districts across the country recommend or even require whole-class or targeted small-group homogeneous instruction. Research evidence demonstrates that the practice of fixed-ability grouping based on student performance—either in whole classrooms or in small groups—is not associated with increased student achievement (Ellis, 2008; Hattie, 2009; Yeh, 2019) and that students arranged into heterogeneous learning groups actually learn more and achieve at higher rates (Boaler, 2006).

 In this book, you will find that we recommend pairs or small-group work for every task-based lesson. In preparing for the task work, consider taking Em's advice to prepare student groups strategically.

To learn more about grouping students using mixed-ability strengths, check out the following resources:

CHECK THESE OUT

Strengths-Based Teaching and Learning in Mathematics: 5 Teaching Turnarounds for Grades K–6   Kobett and Karp (2020)
Reimagining the Mathematics Classroom: Creating and Sustaining Productive Learning Environments   Yeh et al. (2017)

SETTING NORMS FOR TASK-BASED WORK

The task-based lessons in this book require students to work positively and collaboratively to solve problems, share their own ideas, listen to the ideas of others, and offer thoughtful and helpful feedback that moves mathematical understanding forward. But the classroom must be set up for these behaviors to occur. Classroom and task-based norms are the behavioral and cognitive expectations stated using positive strengths-based language that promote equitable student participation. When teachers establish norms for task-based work, they help students to learn how to work together productively to engage in the mathematics. First, many teachers find it helpful to co-construct the norms with the students to develop ownership. Teachers can do this by asking students, "What does it look like and sound like when your group is working together to solve problems?" Second, it is essential to strategically teach the norms to the students and allow them to reflect on how they are enacting those norms each day, even sometimes discussing or role-playing counterexamples. Some of the norms can be used universally for group work when students are sharing their ideas, while other norms are specific to mathematics task work. Note how the Mathematics Task Norms in Figure 2.6 reflect the SMPs.

Figure 2.6 Developing Group and Task Norms

Universal Group Norms	Mathematics Task Norms
• We respect everyone's learning in our class. We do this by praising each other and helping each other. • We ask questions to understand ideas. • We take turns when sharing ideas. • We encourage each other to keep going by saying, "You can do this!" • We make sure that everyone in the group participates. • We listen to each other's ideas and give eye contact when they are explaining their ideas. • We appreciate all ideas, even if we don't agree. • We ask for help when we don't understand something. • We keep trying, even when we make mistakes.	• We talk about our mathematical thinking. • We persevere when solving problems. • We use multiple representations to show our understanding. • We use multiple strategies to solve problems. • We use mathematics vocabulary. • We critique each other's thinking by asking, "How did you decide on that strategy?" • We ask questions to make sense of the math we are learning. We say, "Why did you _____?" and "How did you _____?"

 The tasks in this book will invite students to work in pairs or small groups. If your students don't regularly work together to solve problems or discuss their thinking in small and whole-class discussions, you may find it beneficial to develop the norms before introducing the task-based work and provide ample opportunities for students to reflect on how they used and applied the norms while solving problems.

To learn more about developing and using norms in your mathematics class, check out the following resources:

CHECK THESE OUT

Setting Up Positive Norms in Math Class www.youcubed.org/wp-content/uploads/Positive-Classroom-Norms2.pdf Boaler (2014)
Co-Creating Classroom Norms With Students illustrativemathematics.blog/2019/08/02/co-creating-classroom-norms-with-students/ Illustrative Math (2019)
"Norms and Mathematical Proficiency," *Teaching Children Mathematics* Kasberg and Frye (2013)

When we attend to the selection and implementation of tasks, we also need to make sure that we support students' access, promote productive struggle, consider alternate learning environments, and attend to students' strengths. Therefore, we have designated special places within a task's implementation to call attention to ways that you can promote these important teaching moves.

Student Considerations When Planning *Doing-Math Tasks*

Before we leave this chapter where we unpacked the first half of this template, we explore the following important ideas or teaching moves that will enrich students' learning experiences while they are *doing-math tasks*, and these will be highlighted throughout the tasks as appropriate:

- Access and Equity

- Productive Struggle

- Alternate Learning Environments

- Strengths Spotting

ACCESS AND EQUITY

Knowing your students is the first step in providing equitable learning opportunities and access to high-quality mathematics instruction. NCTM's (2014a) Access and Equity Position Statement says:

> Creating, supporting, and sustaining a culture of access and equity require being responsive to students' backgrounds, experiences, cultural perspectives, language, traditions, and knowledge when designing and implementing a mathematics program and assessing its effectiveness. Acknowledging and addressing factors that contribute to differential outcomes among groups of students are critical to ensuring that all students routinely have opportunities to experience high-quality mathematics instruction, learn challenging mathematics content, and receive the support necessary to be successful. Addressing equity and access includes both ensuring that all students attain mathematics proficiency and increasing the numbers of students from all racial, ethnic, linguistic, gender, and socioeconomic groups who attain the highest levels of mathematics achievement. (n.p.)

Without equal opportunities to access strong, thoughtful mathematics instruction, students' access to learning is limited. Students' learning challenges can be the result of our instructional challenges as we grapple with our teaching decisions regarding how to best meet our students' needs. Frankly, our beliefs about our students' learning potential influences how we plan, what we plan, the questions we ask, how and when we assess, and the feedback we provide. Simply put, we enact teaching that reflects what we believe. If we believe that our students possess deficits and will likely struggle, we may prevent them from engaging in high-quality instruction by reducing the rigor of the tasks we select and implement. Perhaps more concerning, we may break down the task into tiny, bite-size pieces in an effort to make the task accessible, which ultimately prevents our students' access to solve problems using deep mathematical thinking and may make it harder for them to put those pieces back together to see the bigger idea on their own. Our instruction needs to focus on teaching the way that students learn. Pedro Noguera powerfully states the following in an interview (Gonzalez, 2018, "Teach the Way Students Learn," para. 2):

> **We enact teaching that reflects what we believe.**

> Class time needs to be work time for kids. It's only when they're working that a teacher can see who's getting it, who's not, who needs more support. We need to move away from a teacher-centered approach and move toward a student-centered approach. Kids learn through experience. Kids learn through mistakes. Kids learn by asking questions, through interaction. If we taught kids the way they actually learn, our classrooms would look very different than they do right now.

Instruction centered on rich, *doing-math tasks* does just that! NCTM defines equity-based mathematics teaching as the "practices that take into account the way(s) mathematics education perpetuates oppressive norms and therefore actively seeks to erase them, so that *all* students can participate meaningfully in mathematics

In this book, the tasks have been designed to promote access and equity for each and every student through designated opportunities marked with a special icon:

learning and create their own mathematical knowledge" (NCTM, 2020b). These teaching practices include reflecting, noticing, and engaging in a classroom community. Reflecting includes teachers' openness to cultures and perspective; awareness of personal culture, beliefs, and prior experiences; and commitment to Culturally Responsive Mathematics Teaching (NCTM, 2014a). Noticing and attending to your students' mathematical thinking promotes their belief that they are mathematical learners, which in turn promotes the positive development of their identities. Further, teachers can foster a positive and productive classroom community by facilitating rich student discourse where students can reason about mathematics and explain their thinking (Chapin & O'Connor, 2012) using home languages (Turner et al., 2013) and multiple forms of representation.

To learn more about providing access and equity for your students, check out the following resources:

CHECK THESE OUT

The Impact of Identity in K–8 Mathematics: Rethinking Equity-Based Practices
Aguirre et al. (2013)

Catalyzing Change in Early Childhood and Elementary Mathematics: Initiating Critical Conversations
NCTM (2020a)

Countering Deficit Myths of Students With Dis/Abilities and Conceptualizing Possibilities: A Culturally Responsive and Relational Approach to Mathematics TODOS Live!
vimeo.com/353856573
Yeh (2019)

PRODUCTIVE STRUGGLE

Productive struggle in learning mathematics is the process of developing a positive, perseverant mindset for both approaching and pursuing solution pathways when solving problems. When students productively struggle through tasks, they apply prior learning to their new mathematical situations and persistently engage in higher-level thinking (Kapur, 2010). This idea of productive struggle is featured prominently in both the SMPs (National Governors Association & Council of Chief State School Officers, 2010) and in the research-based teaching practices featured in *Principles to Actions* (NCTM, 2014b).

The first of the eight SMPs focuses on helping students *make sense of problems and persevere while solving them* (National Governors Association & Council of Chief State School Officers, 2010). Students' opportunities to experience productive struggle depend, in large measure, on a teacher's instructional decisions. Students cannot develop the ability to productively struggle if they are not presented with consistent opportunities to engage thoughtfully, and yes, productively, when they are learning mathematics. Teachers cannot force students to productively struggle, but they *can* cultivate an environment that nurtures students' stamina and builds their tolerance for pushing through that uncomfortable feeling that materializes between not understanding and understanding. Facilitating productive struggle can be challenging for teachers as they may have been taught to "rescue" children, and this approach that encourages struggle initially may feel indifferent or uncaring. With a few simple moves, you can be on your way to developing a flourishing mathematics classroom environment where students routinely engage in productive struggle as they grow to greater independence (Figure 2.7).

Figure 2.7 Facilitating Productive Struggle

Productive Struggle Move	What Teachers Do	What Students Do
Provide a safe environment.	Co-create group norms with students.	Students ask questions to the teacher and one another, share ideas with their classmates as they solve problems, and encourage each other to try out new ideas and approaches.
Use *doing-math tasks*.	Select and adapt tasks that encourage multiple solution pathways.	Students don't immediately know how to solve a problem or task. They try different strategies and revisit the task to make sense of the problem.
Offer flexible thinking opportunities.	Encourage multiple solution pathways and the use of multiple representations. Ask questions like, "What might be another way to solve this task?" and "What other representations will help you understand this problem?"	Students use multiple strategies flexibly to solve problems. They select and apply representations by attending to the types of problems in the task.
Create a strengths-based classroom.	Celebrate students' strengths and successes.	Students know their strengths and use these assets to solve problems and support their classmates in the problem-solving process.

The opposite of the disposition of productive struggle is one of destructive struggle, which is when students are in a negative cycle and can't move past their frustration or through feelings of helplessness. Some indicators that students are in a destructive cycle include the following:

- Disengaging in the task by not participating or physically showing signs of giving up

- Choosing an alternate activity that is off task

- Retreating and getting quiet, avoiding the teacher's gaze

- Expressing frustration or anger at the task, other students, or the teacher

Your role as a teacher is critical for turning this destructive cycle around as you "greatly influence how students perceive and approach struggle in the mathematics classroom" (NCTM, 2014b, p. 50). Instead, select and adapt tasks that will target that special space of harmonious balance where students are free to persevere and take risks—a place where they want to work hard. Support your students' move away from any form of destructive struggle and instead consistently point to productive struggle by trying these teacher moves (Figure 2.8).

Figure 2.8 Language to Move Students Toward Productive Struggle

Moving Students Toward Productive Struggle →	
Acknowledge your students' feelings while they are learning mathematics.	"I see you are not sure where to start. What have you thought about trying?"
Name perseverance when it is observed.	"I see you tried three strategies to solve the task, Jenna. You are showing perseverance and flexibility by trying new strategies."
Promote collaboration among students.	"I see you are having trouble getting started on this task. Who would you like to work with?"
Explain that being good in mathematics is about problem solving rather than getting answers quickly.	"I know we sometimes feel like we should solve problems quickly, but today we are going to spend the whole class talking about one task!"
Encourage student discourse.	"Talk with a partner about your ideas for getting started on this task. Then each pair will share their ideas in a *Pair-to-Pair Share.*"
Ask students to share stories of perseverance in other contexts (e.g., video games, athletics).	"Tell about a time when you felt frustrated and then you persevered to accomplish your goal."

 In this book, the tasks and lessons have been designed to promote productive struggle. Look for the ✏ icon to promote opportunities for your students to engage in productive struggle.

To learn more about productive struggle, check out the following resources:

🔍 CHECK THESE OUT

Productive Math Struggle: A 6-Point Action Plan for Fostering Perseverance SanGiovanni et al. (2020)
"Productive Struggle for All: Differentiated Instruction," *Mathematics Teaching in the Middle School* Lynch et al. (2018)
"Productive Struggle in Action," *Mathematics Teacher: Learning and Teaching PK–12* Baker et al. (2020)

ALTERNATE LEARNING ENVIRONMENTS

We recognize, understand, and support the fact that learning happens in many environments including outside of the traditional classroom. Particular events such as the COVID-19 pandemic that impacted the world in 2020, weather conditions, individual student medical or mental health conditions, religious observations, or other issues may prevent students from learning inside a classroom. Instruction may occur in online formats, in remote settings, through one-on-one teaching, or through small-group instruction in a nontraditional location. We also recognize that when schools and families co-create learning opportunities, students consistently benefit.

To find resources that support mathematics instruction in alternate learning environments, check out the following resources:

When schools and families co-create learning opportunities, students consistently benefit.

In this book, the alternate learning environment suggestions are noted with a special icon:

🔍 CHECK THESE OUT

Moving Forward: Mathematics Learning in the Era of COVID-19 www.nctm.org/uploadedFiles/Research_and_Advocacy/NCTM_NCSM_Moving_Forward.pdf NCTM and NCSM (2020)
The Math Learning Center: Free mathematics applications for virtual manipulatives, including number lines, clocks, geoboards, base ten materials www.mathlearningcenter.org/resources/apps
Pear Deck: A communication platform used to provide individual feedback to students www.peardeck.com/remote-learning
Google Classroom: A full suite of collaborative documents, slides, whiteboard tools classroom.google.com/

STRENGTHS SPOTTING

When teachers promote their students' strengths, both teachers and students flourish. All students possess strengths in mathematics and often showcase those strengths when they get excited about an idea they have or a strategy they are trying. To best identify students' strengths, we must first look for strengths within the individual student, not in comparison to other students. Students demonstrate their strengths when they persevere through a task, use manipulatives or create representations to show their mathematical understanding, communicate their ideas, seek conceptual understanding by asking "why" when learning mathematics, and collaborate with classmates to solve problems. Teachers recognize and cultivate students' strengths by carefully observing and listening to their students as they make sense of the mathematics they are learning.

There are four steps for strengths spotting, as seen in Figure 2.9.

Figure 2.9 Four Steps for Strengths Spotting

1. Notice the strength.	Teachers plan to recognize their students' strengths. They observe and record the strengths they see. Some teachers use checklists, and others use journals.
2. Name the strength.	Teachers articulate the strength to the student by first recognizing the strength when they see students demonstrating it. For example, teachers might say, "I see Imani is using a number line to compare decimal values." Second, they state, "She is using her mathematics representation strength."
3. Translate the strength.	Teachers explain how the strength helps students learn mathematics. Teachers might say, "Imani is showing a mathematics strength by using a representation to make sense of the value of decimals."
4. Appreciate and value the strength.	Teachers can appreciate the strength or they can call on other students to explain how the students' strength is valuable to them. For example, a teacher might say, "When I see K-Shaud persevering through the task, I am encouraged to keep persevering too. Thank you, K-Shaud, for helping me want to keep trying."

 In this book, the Strengths Spotting suggestions are noted with a special icon:

To learn more about strengths spotting, check out the following resources:

CHECK THESE OUT

Strengths-Based Teaching and Learning in Mathematics: 5 Teaching Turnarounds for Grades K–6

Kobett and Karp (2020)

"Assembling the Puzzle of Students' Mathematical Strengths," *Mathematics Teaching in the Middle School*

White et al. (2018)

"Supporting Teacher Noticing of Students' Mathematical Strengths," *Mathematics Teacher Educator*

Jilk (2016)

Summing Up

Preparing to teach a *doing-math task* requires time and thought to effectively plan. In this chapter, we explored the mathematics process standards, mathematical practices, vocabulary, materials, and the task preparation needed to teach the tasks located in Chapters 4–16. In Chapter 3, we turn our attention to implementing the task.

Professional Learning/Discussion Questions

Read and discuss the following questions with your grade-level teaching team or with teams across multiple grade levels.

- Which of these planning components of the task template are most challenging for you? Why?

- What are some of the shifts in your practice that you feel you need to make? What do you already have in place?

- Why is preplanning so important to this work? What would you need to add to your planning process that you haven't tried before?

- How do you anticipate when your students might be engaged in productive struggle within a task?

- What do you notice about your students when they are engaged in math tasks?

- As you think about preparing to implement a task, how will you use strengths spotting to prepare for your class to engage in the task?

Implementing a *Doing-Math Task*-Based Lesson

In this chapter we move from planning to action, from thinking about and selecting *doing-math tasks* to implementation. We will use this chapter to elaborate on the second half of the task template—Launch, Facilitate, Close, and Reflection—which is all about instructional actions. By the end of this chapter you will

- learn how to launch a *doing-math task*-based lesson by capturing the students' attention, and inviting and affording access for all students;

- understand the dynamics of lesson facilitation by focusing on grouping practices, discourse moves, purposeful questioning, and formative assessment;

- explore the most important elements of task lesson close; and

- consider the importance and significance of recording post-task notes and reflecting about teaching *doing-math tasks*.

Launching a Mathematics Lesson

The lesson launch is very much like those first few paragraphs or pages of a new book you open. The words either grab your attention and summon you to turn the page to learn more, or, sadly, they cause you to abandon the book in disinterest. The *doing-mathematics* task lesson launch is no different! The way teachers choose to launch the task sets the lesson in motion and, in many ways, establishes the tone for the entire experience. From the first powerful moment, "students make conscious or unconscious decisions about whether they will engage in a lesson" (Kobett et al., 2018, p. 136). Additionally, an effective task launch promotes access and equity for students because it helps students engage in tasks from multiple entry points, capitalizing on both their prior knowledge and funds of knowledge about students' background to make sense of what they are experiencing. Leveraging the launch requires careful attention to what researchers Jackson et al. (2012) identify as the four critical elements that should be considered when launching rich mathematics tasks:

1. **Discuss the key contextual features:** Contexts that are unfamiliar to students create barriers because the effort they extend trying to understand the context overworks their working memory (Sweller, 1988). This overextension leaves the students less cognitive space to solve the task.

2. **Discuss the key mathematical ideas:** Students need to know and understand the learning goal for the task. Warning! This element does not mean that the task is unpacked with so much detail that students do not have to think to solve the task. Rather, teachers need to be able to "discuss the key mathematical ideas without hinting at particular methods or procedures that should be used to solve the task" (Jackson et al., 2012, p. 27).

3. **Develop common language to describe the key features:** Teachers carry out this element by asking open-ended questions that encourage students to talk about the task rather than the teacher telling students about the task. Soliciting students' ideas and posing questions about the task support students in developing common language about the task, mathematical ideas, and potential solution pathways.

4. **Maintain the cognitive demand:** A task's degree of cognitive demand was described in detail in Chapter 1 as the level of effort, or mathematical rigor, that students use to solve the problem (Stein et al., 2000). Teachers can do a lot to promote the cognitive intensity of the task in the Launch portion of the lesson by prompting students to generate and use multiple solution pathways. In turn, teachers must be careful not to mandate or promote a particular method because that "robs them [students] of the opportunity to develop mathematical understanding as they generate their own solution methods and representations" (Jackson et al., 2012, p. 28).

You can launch tasks using a variety of techniques that captivate students. The Launch techniques in this book all share one common theme—they promote students' critical thinking about the problems they are solving. The tasks students encounter are all problems to be solved within particular contexts that connect to specific topics and standards within the grade level. The launch should support, not detract from, the goal and should always allow the mathematics task to be problematic. Hiebert et al. (1996) explain that "allowing the subject to be problematic means allowing students to wonder why things are, to inquire, to search for solutions, and to resolve incongruities. It means that both the curriculum and instruction should begin with problems, dilemmas, and questions for students" (p. 12). The launch builds curiosity and a desire to solve rich, thoughtful problems.

> " **Launches should captivate students.** "

LAUNCH TECHNIQUES

The following launch techniques are used in Chapters 4 through 16:

- Discussion-Centered Launches

- See, Think, and Wonder

- Notice and Wonder

- Which One Doesn't Belong?

- Games

- Children's Literature

We'll introduce you to them here, briefly, in order to familiarize you with them and to give you a hint of what you will see in the tasks. You may be familiar with some of these launches and incorporate them already. If some are new, they can be added to your toolbox!

Discussion-Centered Launches

Students need opportunities to process and discuss their ideas before moving toward solution pathways. Every task-based lesson in this book provides multiple opportunities for students to engage in mathematical discussions in pairs, small groups, and a whole group using a variety of approaches. These discussion techniques are described in detail and may include *Turn and Talk, Turn and Learn, Think-Pair-Share,* and *Pair-to-Pair Share.* Explanations and examples of these techniques are described in detail in this chapter in Figure 3.1, Grouping Structures and Techniques.

See, Think, and Wonder

The See, Think, and Wonder launch ignites students' curiosity, engages them in an inquiry cycle, and prepares them to ask mathematical questions (Ritchhart et al., 2011) using visual images, graphs, and video. Students are invited to recall what they think about the content before being directed to engage with the math in particular ways. This launch technique is particularly helpful for students who are learning mathematics in a language that is different from their home language. See Chapter 4, Task 4, p. 58 for an example of See, Think, and Wonder in action.

To learn more about See, Think, and Wonder, check out the following resource:

 CHECK THIS OUT

Making Thinking Visible: How to Promote Engagement, Understanding, and Independence for All Learners
Ritchhart et al. (2011)

Notice and Wonder

Similar to See, Think, and Wonder, the Notice and Wonder launch protocol (Math Forum, 2015) also summons students' observations and ideas about what they are seeing. Using visual images, videos, graphs, and/or manipulatives, students are prompted with "What do you notice?" and "What do you wonder?" Teachers record what students notice and wonder on a chart and then use the students' ideas to present the task. See Chapter 6, Task 8, p. 75 for an example of what Notice and Wonder looks like within a task.

To learn more about the Notice and Wonder launch protocol, check out the following resources:

 CHECK THESE OUT

"Beginning to Problem Solve With 'I Notice and Wonder'"
www.nctm.org/Classroom-Resources/Problems-of-the-Week/I-Notice-IWonder/
Math Forum (2015)

"Capturing Mathematical Curiosity With Notice and Wonder," *Mathematics Teaching in the Middle School*
Rumack and Huinker (2019)

Which One Doesn't Belong?

Fashioned after the old *Sesame Street* song and corresponding visual activity, "Some of these things are not like each other, some of these things are kind of the same" (Raposo & Stone, 1972), these lesson launches ask students to look at four visual images, representations, numerals, or expressions to name and explain how one is different from the others (or how some are the same). These launches work best when each of the options can be identified as a viable response for the one that does not belong. These launches invite students to construct viable arguments, use mathematical vocabulary, incorporate prior knowledge, and make connections. See Chapter 13, Task 43, p. 216 for an example of this launch technique.

To learn more about the Which One Doesn't Belong? launch technique, check out the resources in the sidebar.

CHECK THESE OUT

Which One Doesn't Belong? wodb.ca/ Barousa (2013)
Which One Doesn't Belong? A Shapes Book and Teacher's Guide Danielson (2016)

Games

Games immediately engage students collaboratively and provide opportunities that support students in making conjectures and predictions, and developing conceptual understandings. Games can also promote equity because *all* students learn and play the game at the same time, providing a common, shared, collaborative experience in their learning community. Once students have played the game, teachers can extend the experience by placing games in centers and sending them home with children to play with their families. See Chapter 9, Task 22, p. 131 for an example of this launch technique.

To learn more about using Games as a launch technique, check out the following resources:

CHECK THESE OUT

Games for Math: Playful Ways to Help Your Child Learn Math: From Kindergarten to Third Grade Kaye (2012)
Math Games & Activities From Around the World Zaslavsky (1998)
Math Fact Fluency: 60+ Games and Assessment Tools to Support Learning and Retention Bay-Williams and Kling (2019)
Illuminations NCTM (n.d.)

Literature

Similar to Games, launching math lessons with Children's Literature creates a shared experience for teachers and students (Furner, 2018). Children's Literature draws students into real life or fanciful contexts that introduce new mathematics concepts or reinforce previously learned mathematics concepts. This interdisciplinary approach supports students in making connections to their personal experiences and using critical thinking skills to solve problems and remember the story long after it has been read (Murphy, 2000). We recognize that not all teachers have access to literature. Therefore, we offer literature suggestions as alternative options for lesson launches. See Chapter 14, Task 47, p. 235 for an example of this launch technique.

To learn more about using Children's Literature as a launch technique, check out the resources in the sidebar.

CHECK THESE OUT

Math and Literature: Grades K–1 Burns and Sheffield (2004)
Exploring Math Through Literature, Pre-K–8 NCTM (2017a)

Facilitating the Lesson

Facilitating a *doing-math task*-based lesson is so much more than merely assigning a task for students to complete! Teachers must weigh and regularly consider many factors as they facilitate a task that will maximize the students' learning opportunities. Throughout each of the tasks provided, you will find suggestions that include considerations of the following:

- How students will be grouped to engage in and solve the task

- How and when students will engage in discourse

- Purposeful questions you may ask that will challenge the students or move them to the next stage

- Classroom-based formative assessment techniques you may use to understand student thinking

> In this book, you are encouraged to facilitate grouping your students in multiple ways (Figure 3.1). Don't be discouraged if initial grouping attempts are challenging. Keep working at it to find flexible grouping structures that work for you and your students!

GROUPING STUDENTS

In Chapter 2, you read about grouping students for task work. We strategically placed the grouping discussion in task preparation because grouping structures that are decided ahead of time avoid grouping decisions that don't capitalize on students' strengths. As noted, students who work in mixed-strengths groups are able to contribute their strengths as a cooperative group to solve problems. As you gain comfort implementing *doing-math tasks*, consider how you can flexibly group your students in pairs or groups of three or four to best support their learning on a particular task. Observe which students work best together, make notes, and adjust the grouping as needed.

Figure 3.1 Grouping Structures and Techniques

Group	When to Use This Grouping Structure	Techniques
Pairs	Pairs work well for engaging students in all launch techniques, brainstorming solution pathways during the Facilitate portion, and revealing solutions during the Close activities.	*Turn and Talk*: Turn and share your ideas with a partner. *Turn and Learn*: Ask your partner a question about their thinking. Be prepared to share your partner's ideas in the whole-group discussion.

Group	When to Use This Grouping Structure	Techniques
Threes	Groups of three can brainstorm ideas during the Launch and Facilitate parts of the lesson. Groups of three can be particularly effective when students are creating multiple representations.	Groups of three can be assigned particular roles within a group. *One Stay, Two Stray:* After a group has worked together to brainstorm ideas or solve a task, the teacher asks one person in the group to stay and the two others to "stray" and join another person who has "stayed" to share ideas. Once new groups are formed, the students share their groups' ideas or answer a new question. This approach can be facilitated by assigning each student a number. For example, number 1 stays, numbers 2 and 3 stray (find another group).
Fours	Two pairs of students share and compare strategies in groups of four.	*Pair-to-Pair Share:* Pairs first share ideas together and then join another pair to share and discuss their ideas. *Pair-to-Pair Interview:* Pairs examine another pair's work and then confer to decide on questions they will ask about the students' work.

Grouping students can be challenging as students build skills in cooperating with one another, explaining their ideas, and listening to one another attentively. For more ideas about establishing group norms, check out pages 19–20 in Chapter 2.

DISCOURSE

A classroom that is rich in mathematical discourse does not happen by accident. It happens with careful and thoughtful planning and the conviction that all students can engage in meaningful communication! The Standards for Mathematical Practice (National Governors Association & Council of Chief State School Officers, 2010), Process Standards (NCTM, 2000), and the Effective Mathematics Teaching Practices (NCTM, 2014b) all emphasize the importance of facilitating student discourse to promote students' mathematical understanding of important concepts and skills (Figure 3.2). Note how the responsibility of this discourse is placed squarely on the shoulders of the students.

> " A classroom rich in mathematical discourse does not happen by accident. "

Figure 3.2 Crosswalk Between Standards for Mathematical Practice, Process Standards, and Effective Mathematics Teaching Practices

Standards for Mathematical Practice (National Governors Association & Council of Chief State School Officers, 2010)	Process Standards (NCTM, 2000)	Effective Mathematics Teaching Practices (NCTM, 2014b)
Construct Viable Arguments ... (pp. 6–7) • Mathematically proficient students understand and use stated assumptions, definitions, and previously established results in constructing arguments. They make conjectures and build a logical progression of statements to explore the truth of their conjectures. They justify their conclusions, communicate them to others, and respond to the arguments of others. • Mathematically proficient students are also able to compare the effectiveness of two plausible arguments, distinguish correct logic or reasoning from that which is flawed, and—if there is a flaw in an argument—explain what it is. • Elementary students can construct arguments using concrete referents such as objects, drawings, diagrams, and actions. Such arguments can make sense and be correct, even though they are not generalized or made formal until later grades.	Communication (pp. 60–63) Students • organize and consolidate their mathematical thinking through communication; • communicate their mathematical thinking coherently and clearly to peers, teachers, and others; • analyze and evaluate the mathematical thinking and strategies of others; and • use the language of mathematics to express mathematical ideas precisely.	Facilitate Meaningful Mathematical Discourse (p. 35) Students • present and explain ideas, reason, and represent to one another in pair, small-group, and whole-class discourse; • listen carefully to and critique the reasoning of peers, using examples to support or counterexamples to refute arguments; • seek to understand the approaches used by peers by asking clarifying questions, trying out others' strategies, and describing the approaches used by others; and • identify how different approaches to solving a task are the same and how they are different.

Without opportunities to question, discuss, share, critique, and defend their ideas, students won't experience the richness and beauty of mathematics. Quite simply, discourse helps students discover and make their own connections to the mathematics they are learning. *Principles to Actions* (NCTM, 2014b) explains this important idea in the teaching practice Facilitate Meaningful Mathematical Discourse: "Effective teaching of mathematics facilitates discourse among students to build shared understanding of mathematical ideas by analyzing and comparing student approaches and arguments" (p. 10). Planning for and teaching in ways that support student discourse requires a clear examination of the teacher and student roles in the classroom. By examining the ratio of teacher-to-student talk, we can get a good idea about the opportunities for discourse that we provide for our students. Students should be discussing their ideas at every phase of the task-based lesson but particularly during the Facilitate component of the lesson as they engage with their partners or small groups to share their ideas. We can all ask a series of questions to better understand how we are promoting discourse in our classrooms.

Are we

- productively grouping students to engage in meaningful discourse?

- highlighting meaningful student discourse when we hear it and see it?

- ensuring that students have time to grapple with mathematical ideas?

- using class norms that promote students as authors of their own mathematical ideas?

- asking questions that promote meaningful discourse?

> Throughout the tasks provided in this book, you should notice many opportunities to engage students in discourse through rich, thoughtful, and strategic questions.

To learn more about providing rich discourse opportunities, check out the following resources:

CHECK THESE OUT

Classroom Discussions: Seeing Math Discourse in Action, Grades K–6 Anderson et al. (2011)
Number Talks: Whole Number Computation, Grades K–5 Parrish (2014)
The 5 Practices in Practice: Successfully Orchestrating Mathematical Discussions in Your Elementary Classroom Smith et al. (2019)
Intentional Talk: How to Structure and Lead Productive Mathematical Discussions Kazemi and Hintz (2014)

PURPOSEFUL QUESTIONS

The kinds of questions that teachers ask during a lesson hold the opportunity to move students' mathematical understanding forward in deep and powerful ways. NCTM's *Principles to Action* (2014b) states that "effective teaching of mathematics uses purposeful questions to assess and advance students' reasoning and sense making about important mathematical ideas and relationships" (p. 35). As teachers develop questions that will prompt students to think deeply about the mathematics they are learning; make conjectures, generalizations, and conclusions; and extend their curiosity about the task (Van de Walle et al., 2019), they should consider the five question types (see Figure 3.3; NCTM, 2014b, 2017c) to ensure that students are engaged in a variety of levels of mathematical thinking. Beyond Bloom's Taxonomy (Bloom, 1956), the five question types in Figure 3.3 illustrate how strategic questioning can move students' learning forward.

Figure 3.3 NCTM's Five Question Types With Examples

Question Type	Teachers Ask These Types of Questions to . . .	Task Examples (with page numbers)
Gathering Information	Elicit procedural information that has right and wrong answers	From Chapter 11, Task 35, p. 179: "Is Tim's conjecture true or false?"
Probing Thinking	Encourage students to demonstrate their reasoning by explaining their ideas and strategies	From Chapter 10, Task 31, p. 163: "How can you prove that they are correct using your counters or base ten blocks?"
Making the Mathematics Visible	Support students to recognize patterns, connect mathematical ideas, and understand the underlying structure of the mathematics content they are learning	From Chapter 8, Task 18, p. 114: "What would be an example of an equation to represent what just happened?"
Encouraging Reflection and Justification	Support students to develop their mathematical arguments and justify their solution pathways with deep explanations and representations	From Chapter 12, Task 36, p. 187: "What did your group discover? Were there any surprises? How do you know which objects are heavier than the others?"
Engage With the Reasoning of Others	Encourage students to construct a viable argument, listen to the reasoning of their peers, and ask questions of their peers	From Chapter 12, Task 37, p. 191: Allow student groups to join another group of students for a group-to-group share. Say, "Share your work. Look at their bears. What do you notice? Did they solve the task the same way you did? Did they solve it differently?" Allow time for students to share their completed work, explain their answers, and ask questions if needed within their new groups.

In this book, some questions have been provided for you in the tasks, but certainly you will decide on other robust and rich questions that you could actually ask in the task lesson. To ensure that you are asking questions from all of the question type categories, consider using Figure 3.3 to plan your questions as you facilitate the task.

To learn more about the importance and use of questioning, check out the following resources:

 CHECK THESE OUT

"Teacher Questioning to Elicit Students' Mathematical Thinking in Elementary School Classrooms," *Journal of Teacher Education*
Franke et al. (2009)

Good Questions: Great Ways to Differentiate Math Instruction in Standards-Based Classrooms
Small (2017)

FORMATIVE ASSESSMENT

Classroom-based formative assessment, also known as formative evaluation (Hattie, 2009), is assessment *for* learning because it centers on collecting information about student understanding while you are teaching and using that information to respond to students in real time, at the moment you are facilitating the task. The formative assessment research that follows provides a powerful reminder of the effects of collecting student responses and data and using them to influence planning and instruction. Formative assessment, when used regularly and strategically, advances student learning.

Wiliam and Thompson (2008) identified the following five research-informed formative assessment strategies that, when implemented, foster student learning.

1. Clarifying and sharing learning intentions and criteria for success

    What: Teachers let students know what they will be learning, how they will be learning it, and how they will know when they are successful using student-friendly language. Sometimes it is appropriate to withhold the exact content of the task to allow students to make their own connections and conjectures. In such cases, it is appropriate to let students know that, for example, they will be solving a problem using multiple representations.

    When: Teachers discuss the learning intentions and success criteria at the beginning of the task and throughout the task.

2. Engineering effective classroom discussions, questions, and learning tasks that elicit evidence of learning

    What: Teachers do this throughout the task's implementation by asking questions, probing student thinking, and asking students to use multiple representations and solution pathways to demonstrate their understanding.

    When: Teachers promote this strategy throughout the Facilitate and Close portion of the lesson.

3. Providing feedback that moves learning forward

    What: Teachers provide feedback that attends, responds, and is crafted to target students' understanding in the moment for individual students, and small and large groups, in an effort to advance students' mathematical thinking.

    When: Teachers provide feedback throughout the Facilitate and Close portions of the task-based lesson.

4. Activating students as owners of their own learning

    What: Teachers bolster their students' mathematical competence and identities by communicating confidence to them regarding their ability to solve tasks. They turn over mathematical authority to their students by asking them to share ideas and lead discussions with peers.

When: Teachers can and should activate student ownership throughout the entire lesson.

5. Activating students as instructional resources for one another

What: Students do this when they support one another by asking questions, offering new or alternative ideas, and sharing their solution pathways.

When: Students need opportunities to serve as a resource while solving tasks together during the Facilitate part of the task-based lesson and in the lesson close when the teacher is helping students strategically bring mathematical ideas together.

> " When used strategically, the Formative 5 Assessments support teachers in collecting important indicators of students' understandings. "

> Throughout the task's Facilitate section, you will find the Formative 5 techniques indicated in bold along with a prompt or question.

Five classroom-based formative assessment techniques also known as the Formative 5, when used strategically throughout a lesson, support teachers in their ongoing collection of important indicators of the mathematical understandings of their students (Fennell et al., 2017). The five techniques are Observations, Interviews, Show Me, Hinge Questions, and Exit Tasks (see Figure 3.4). By using these techniques, the student responses and resulting data enable teachers to provide strategic and thoughtful feedback that moves student learning and instruction forward, as well as helping to guide teacher planning and instructional decision making.

The first three techniques—Observation, Interview, and Show Me—are used often and seamlessly during mathematics task-based work and may be used in the Launch, Facilitate, and Close portions of the task lesson. Teachers will often use a Hinge Question as a task lesson check for understanding/proficiency at a particular point in the lesson. Student responses to these techniques may also inform a decision to drive the lesson in a different direction or point to a different or adapted task. The Exit Task will be implemented less frequently because the students will have already been engaged in a *doing-math task*. Figure 3.4 describes the Formative 5 assessment techniques and provides selected examples of where each technique is used in a task in this book. Each task will feature the use of the Formative 5 techniques. The tasks also refer to a formative assessment tool to record observations and collect student responses from Interview and Show Me prompts.

Figure 3.4 Formative 5 Assessment Techniques

Formative 5 Technique	Description	Task Examples (with page numbers)
Observation	Teachers conduct strategic observation of students while they are working individually, in pairs, or in groups. Typically, students are representing their solution pathways with representations such as manipulatives and mathematical sketches or drawings. Teachers record their observations using observation charts and use observations to ask strategic questions as well as inform their planning and teaching.	From Chapter 9, Task 20, p. 125: • **Observe.** Monitor student progress with the task. Look for the following: • Are students able to write equations that are true? • Are the equations all of the same form? For students who generated many equations using the same format, you may want to emphasize that equations can be true in many different ways. Pair them up with a student who has written a true equation in a different form. • Are students using only one mathematical operation?

Formative 5 Technique	Description	Task Examples (with page numbers)
Interview	Teachers conduct a brief interview when students are working to learn more about their thinking. They ask the kinds of questions that invite students to share their ideas and reasoning. Teachers can collect interview data by recording students' responses on recording sheets (response on interview tools for both individual students and small groups are provided in Appendix B).	From Chapter 10, Task 26, p. 148: **Interview.** Ask the following: • How do you know if Maria or Jamar has more buttons? • Maria has 16 buttons. What do you know about the number 16? • Jamar has 15 buttons. How can you describe 15 using the tens and leftover ones?
Show Me	The Show Me technique is "a performance response by a student or a group of students that extends and often deepens what was observed and what might have been asked in an interview" (Fennell et al., 2017, p. 63). Students most often display their response using some type of representation. Whole-class Show Mes can be conducted using whiteboards and technological tools.	From Chapter 12, Task 36, p. 186: **Show Me.** "Can you show me how we could use the balls to help answer this question?" Allow time for students to discuss either with a partner or as table groups. Allow students to share their responses with the whole group. (Students should be able to demonstrate during the **Show Me**) and articulate that even though the Wiffle ball is larger in size, it is lighter than the golf ball so Mark is incorrect.
Hinge Question	The hinge question is a particular type of question that is used at a strategic or pivotal moment in the lesson that will likely assess student progress with the lesson and drive the lesson in one direction or another (Wiliam, 2011).	From Chapter 14, Task 45, p. 230: **Hinge Question.** Ask, "In what ways do graphs help us see information in a quicker and easier way?"
Exit Task	The exit task is a "capstone problem or task that captures the major focus of the lesson for that day or perhaps several days and provides a sampling of student performance" (Fennell et al., 2017, p. 109). Unlike an exit ticket, the exit task is "meatier" (more demanding) and provides a full range of student thinking about the mathematics topic, or perhaps extends the task-based lesson of the day.	From Chapter 12, Task 41, p. 208: **Exit Task.** Show students the stapler from the beginning of the task and remind them how long the stapler was in large paper clips. Give each student a centimeter cube. "Use what you know about the size of a unit and the number of units it takes to measure an object to tell me if you think we will need more, fewer, or the same amount of centimeter cubes as large paper clips to measure the length of the stapler. Explain your answer using words, pictures, or both."

Although some Formative 5 techniques are included for you in each task, we encourage you to construct your own prompts for each of the Formative 5 techniques. Use the tool templates provided at the companion website (**resources.corwin.com/classroomreadymath/K-1**) to formulate your own Formative 5 prompts.

To learn more about the Formative 5 techniques and formative assessment, check out the following resources:

 CHECK THESE OUT

The Formative 5: Everyday Assessment Techniques for Every Math Classroom Fennell et al. (2017)
"Classroom-Based Formative Assessments: Guiding Teaching and Learning." In C. Suurtamm and A. Roth McDuffie (Eds.) *Annual Perspectives in Mathematics Education* (pp. 51–62) Fennell et al. (2015)
A Fresh Look at Formative Assessment in Mathematics Teaching Silver and Mills (2018)
Mathematics Formative Assessment, Volume 1: 75 Practical Strategies for Linking Assessment, Instruction, and Learning Keeley and Tobey (2011)

Close the Lesson: Make the Mathematics Visible

Ask any teacher, instructional coach, or leader and you will likely find that they seemingly all agree that lesson closure is a critical part of lesson planning yet it often gets pushed to the side in favor of additional instructional time (Ganske, 2017). Not providing lesson closure or even rushing its implementation, however, means that students lose opportunities to make connections between and among mathematical representations, discuss solution pathways, consider which strategies are most efficient, and leave with important mathematical understandings, as well as a sense of what's happening next. Effective and thoughtful close activities improve student retention of material (Pollock, 2007), particularly those close activities that ask students to think, respond, write, and discuss concepts (Cavanaugh et al., 1996). Consider the important role that a Hinge Question or Exit Task may play within a lesson's close. In general, close activities deepen student thinking and offer opportunities for students to make sense of the mathematics they are learning, see how peers are thinking about the task, make generalizations, identify patterns, clarify emerging mathematical understandings, and advance their thinking as they consider more complex mathematical ideas. Such close activities also provide major signals as to your next steps with regard to planning and instruction.

There may be two parts to a lesson's close. The first part of the close, which is the term we are using in our task-based lesson template, provides opportunities for students to share their solution pathways and review classmates' strategies.

Once the students have had opportunities to review one another's work, the teacher makes the mathematics from the task visible by orchestrating productive discussions (Smith & Stein, 2018) using students' strategies and solution pathways from the task. During the close of the task lesson, the teacher focuses on the last three practices proposed by Smith and Stein. The first two practices, anticipating students' solutions to the mathematics task (referenced in Chapter 1) and monitoring (described in the Facilitation portion of this chapter under formative assessment, particularly the use of observation) inform the teacher's decision making in the close. The final three practices, selecting, sequencing, and connecting, are key practices that may be implemented during the close. Consider the following:

- **Selecting:** After observing and monitoring student work, teachers select the student work they will highlight during the close. They choose student work that represents a variety of solution strategies, ideas, and representations.

- **Sequencing:** The teacher then strategically sequences the students' work to share with the class as it aligns to the mathematics goal for the task. For example, teachers might have three different students share their work reflecting a variety of representations moving from the concrete to the abstract. Or, the teacher may sequence the sharing by highlighting the most common strategy to the most unusual.

- **Connecting:** Finally, the teacher connects the students' approaches, unpacks the underlying mathematics, and connects the students' strategies and solution pathways back to the mathematics goal of the task and task lesson.

> In this book, the close activities for the task lessons are designed to be robust and full of varied representations and rich student discourse that make the mathematics from the task visible to the students.

As you plan to implement a task lesson's close activities, consider the amount of time that you and the students will need to participate in deep discussions that unpack the mathematics of the task. Students and teachers may be asked to engage in one or more of the following close activities that prompt students to review the mathematics concepts, strategies, and ideas that were developed during the task lesson's work (Figure 3.5).

Figure 3.5 Task-Based Lesson Close Techniques, Descriptions, and Examples

Close Technique	Description	Where Can I Find Examples?
Open Gallery Walk	Student work is displayed around the classroom. Students are asked to walk around the classroom to look at their peers' strategies and solutions. The teacher uses the student work to strategically discuss the mathematics goal for the lesson, making the mathematics learning visible.	See • Chapter 9, Task 23, p. 136
Something Similar and Something Different Gallery Walk	Student work is displayed around the classroom. Students are asked to walk around the classroom to find peers' work that is similar to and different from their own work. Students place a sticky note or a colored dot on the work to indicate that the work is similar (e.g., pink dot) or different (e.g., green dot). The teacher begins the discussion by reviewing the students' sticky dots, asking the students to share how the work is similar and different regarding the strategies and solution pathways used and then strategically pointing to the mathematics goal for the lesson.	See • Chapter 14, Task 45, p. 229 • Chapter 15, Task 50, p. 254

(Continued)

(*Continued*)

Close Technique	Description	Where Can I Find Examples?
Notice and Wonder Gallery Walk	In this Gallery Walk, students write their notices and wonders on sticky notes and place them on the student work. The teacher begins the discussion by reviewing the students' sticky notes, asking the students to share what they noticed and wondered about the strategies and solution pathways, and then strategically pointing to the mathematics goal for the lesson.	See • Chapter 11, Task 34, p. 178 • Chapter 15, Task 49, p. 249
Select, Sequence, and Connect	During the Facilitate portion of the lesson, teachers look for student thinking and identify specific ideas that will be shared in the lesson close. Teachers ask specific and strategic questions to highlight students' strengths and reasoning about the mathematical ideas. Then, teachers gather students together for a whole-group discussion. They strategically select student work to display and then ask other students to share their strategies and representations. In addition, they "focus students' attention on the structure or essential features of mathematical ideas" (NCTM, 2014b, p. 24) to make the mathematics visible. Teachers also facilitate discussions about efficient and novel strategies by encouraging and coaching students to share solution pathways and make connections among representations.	See • Chapter 7, Task 13, p. 94 • Chapter 10, Task 26, p. 148

To learn more about close techniques and strategies, check out the following resources:

 CHECK THESE OUT

Five Practices for Orchestrating Productive Mathematics Discussions (2nd ed.)
Smith and Stein (2018)

The 5 Practices in Practice: Successfully Orchestrating Mathematics Discussions in your Elementary Classroom
Smith et al. (2019)

Talk Moves: A Teacher's Guide for Using Classroom Discussions in Math
Chapin et al. (2013)

Lesson Reflection

Teacher reflection is key to sustained and continued success. Consistent reflection helps teachers understand the *why* behind the events that happen in their classroom and promotes professional growth (Danielson & McGreal, 2000; Dewey, 1933). Without reflection about how our teaching decisions connect to student understandings, we are left with teaching through imitation rather than through strategic intentionality (Lortie, 1975). This intentional and strategic approach capitalizes on both the mathematics content and our students' strengths and needs. You can support your own reflection process by recording your thoughts in a journal, asking students to give you feedback, or asking a colleague to observe or listen to the story of your lesson. When teachers reflect on their lessons, they learn more about their teaching practice and make connections about how their teaching decisions influence students' learning (Danielson, 2008).

> " Without reflection, we are left with teaching through imitation rather than teaching through strategic intentionality. "

In this book, we include, at the end of each task, an opportunity for you to provide post-task notes regarding your reflective comments as well as thoughts related to your next steps (planning and instructionally).

To learn more about teacher reflection, check out the following resources:

🔍 CHECK THESE OUT

Ten Ways to Be a More Reflective Teacher

Terry Heick (2019), TeachThought.com

www.teachthought.com/pedagogy/reflective-teacher-reflective-teaching/

Math Workshop: Five Steps to Implementing Guided Math, Learning Stations, Reflection, and More

Lempp (2017)

Using This Book to Get Started With *Doing-Math* Tasks

This book's collection of *doing-math tasks* is organized by mathematics topics for kindergarten and first grade into chapters. Each chapter includes a chapter opener that describes the mathematics standards and topics per task, and considerations for anticipating student thinking around those standards, as well as anticipating the task's implementation. As you consider the mathematical topics, it is very helpful to anticipate student thinking to make the most out of your task implementation.

ANTICIPATING STUDENT THINKING

Before setting out on a run, cross country runners visualize the entire race course, anticipating every twist, turn, and hill. They don't just think about the race, they embody it. While they visualize, they also imagine their response to each of the race course elements. They know that there will still be surprises on the course, but their prior anticipation allows them to respond to expected challenges with expert, almost automatic responses that provide space to handle those unexpected, unanticipated challenges.

Similarly, when teachers anticipate how students will engage in a task, they need to imagine the lesson. Where are the hills that need momentum or the sharp turns that need a slower pace? By mixing your knowledge of your students with your teaching experience, you are prepared for and can respond to students with thoughtful questions, prompts, and probes that will advance your students' thinking. While students will still catch us by surprise sometimes with their wonderful and unique thinking, the more prepared we are, the more likely we can respond with appreciation and the less likely we are to shut down student thinking. This approach also helps us avoid correcting students before they have had a chance to engage in sense making or convince us with logical arguments. It is natural for teachers to focus on what students don't know, and frankly, we worry about what they don't yet understand. However, a deficit mindset about naturally developing mathematical understanding can harm students because we may respond with corrective feedback rather than with curiosity or questions. We purposely do not use the word *misconception*, instead focusing only on student thinking, without judgment, so that we can gracefully advance their thinking and understanding.

You can use the following teaching moves to anticipate student thinking:

1. **Do the task!** Engaging in the task first will allow you to think about the nuances of the task components, including the mathematics, context, vocabulary, strategies, and solution pathways.

2. **Anticipate *all* student responses.** Write down the ways that your students may respond, and write questions that you will ask in reply to that thinking.

3. Think about students' strengths. Leverage their strengths as well as their challenges.

Additional support, including student pages, can be found on the book's companion site.

As you consider your next steps instructionally, you may have some technical questions.

Q: Where should I start?

A: We suggest that you begin with a mathematics standard that is particularly important to you and select a task to try first. If possible, pick a teaching partner to plan with and check back with one another to share your experiences!

Q: Where do I find the student pages?

A: The student pages can be found at **resources.corwin.com/classroomreadymath/K-1**

Q: How much time do I need to teach the tasks?

A: We recognize that the time allotted to teach mathematics varies greatly across this nation. As you read the task lesson, note that the Launch may require up to 10 minutes, the Facilitate portion will likely require 30 minutes as student groups work together to formulate strategies and solutions, and the Close will require at least 10–15 minutes, and sometimes much more to fully unpack the mathematical understanding. Of course, these are all estimates.

To learn more about anticipating student thinking, check out the following resources:

 CHECK THESE OUT

Every Math Learner, Grades K-5: A Doable Approach to Teaching With Learning Differences in Mind
Smith (2017)

"Three Strategies for Opening Curriculum Spaces"
Drake et al. (2015)

Summing Up

In this chapter, we considered important components of task implementation such as the lesson launch, lesson facilitation, and lesson close. Task lesson launches provide opportunities for students to attend to and demonstrate curiosity about the task. As teachers facilitate lessons, they focus on grouping practices and student discourse opportunities, pose purposeful and thoughtful questions, and use formative assessment to assess student understanding. The task lesson close is particularly important when implementing a *doing-math task* as teachers strategically use students' solution pathways to make the mathematics visible to students.

Professional Learning/Discussion Questions

Read and discuss the following questions with your grade-level teaching team or with teams across multiple grade levels.

- Why is a task lesson launch important?

- Which of the task lesson launch techniques have you already tried?

- What is the role of questioning in a task lesson?

- How does formative assessment during the lesson help you support students' learning?

- Why is the close a critical step in teaching a mathematics task lesson?

Counting and Cardinality

Counting and Writing Numbers

GETTING STARTED

TASK 1: KINDERGARTEN: QUICK COUNTS

Count forward beginning from a given number within the known sequence (instead of having to begin at 1).

Count to answer "how many?" questions about as many as 20 things arranged in a line, a rectangular array, or a circle, or as many as 10 things in a scattered configuration; given a number from 1–20, count out that many objects.

TASK 2: KINDERGARTEN: RACE TO 100

Count to 100 by ones and by tens.

TASK 3: KINDERGARTEN: IT'S A MATCH!

Write numbers from 0 to 20. Represent a number of objects with a written numeral 0–20 (with 0 representing a count of no objects).

TASK 4: KINDERGARTEN: WHAT'S IN A DAY?

Write numbers from 0 to 20. Represent a number of objects with a written numeral 0–20 (with 0 representing a count of no objects).

~~~~~~

Anticipating Student Thinking: And so it begins! Mathematics at the kindergarten level is both fun and foundational. This book's beginning task chapter presents four tasks that focus on the importance of counting. Students will be actively engaged in tasks where they will be counting forward from a beginning number, counting all, counting by ones and tens, and writing and representing numbers from 0 to 20. Each of the chapter's tasks will engage your students as they use a variety of representations to both develop their understanding and assist them as they count and write numbers. Each of the task lessons involves classroom-based formative assessment techniques to monitor student progress and provide feedback to you as you anticipate student thinking and consider your next steps instructionally. Let's get to it!

THINK ABOUT IT

As you prepare for the implementation of the chapter's tasks, think about how you will ensure that all of your students have access to each task's activities within the Launch, Facilitate, and Close components of the task lesson. Additionally, be on the lookout for student strengths regarding their comfort and understanding with counting, as well as representing and writing numbers.

Mathematics Standards

- Count forward beginning from a given number within the known sequence (instead of having to begin at 1).

- Count to answer "how many?" questions about as many as 20 things arranged in a line, a rectangular array, or a circle, or as many as 10 things in a scattered configuration; given a number from 1–20, count out that many objects.

Mathematical Practices

- Attend to precision.
- Look for and express regularity in repeated reasoning.

Vocabulary

- count on
- count all
- next

Materials

- Counting Cards student pages (prepare multiple sets so each pair can try new cards as time allows)
- Teacher Model Counting Card
- blocks or counters in paper bags
- blank paper or marker board for recording

Task 1
Quick Counts

Connect counting on to counting all

TASK

Quick Counts

Johnna and Terrence are practicing counting things (see Figure 4.1). Johnna started counting hearts. When she got to 7, Terrence said, "I'll finish!" Help Terrence count on from where Johnna left off.

Figure 4.1 Terrence Counts Hearts

Source: child clipart: pixabay.com; heart clipart: publicdomainpictures.net

How many hearts are there?

TASK PREPARATION

- Use this task to facilitate the transition from counting all to counting on from a given number.

- Consider how you will organize students into heterogeneous pairs.

ACCESS AND EQUITY

Research shows that students in heterogeneous groups learn more and have higher achievement. When students can hear diverse perspectives and engage in dialogue with their peers, they have opportunities to engage with the content at deeper levels.

- Print, cut, and fold Student Counting Cards. Consider laminating these so students can mark them up with dry-erase markers.

- Prepare a paper bag with 10 blocks or counters inside (for the Hinge Question at the end of the Close phase of the lesson). Write the numeral 10 clearly on the outside of the bag.

LAUNCH

1. Facilitate a See, Think, and Wonder using the image from the task (See Figure 4.2).

Figure 4.2 Terrence Counts Hearts

Source: child clipart: pixabay.com; heart clipart: publicdomainpictures.net

2. Have students *Turn and Talk* to a partner, then have students share out their ideas with the class.

3. Record students' wonderings on the board.

4. Tell students the story about Johnna and Terrence counting hearts.

5. Indicate the part of the image with the number. Ask, "The part Johnna already counted is covered up. How many did Johnna count already? How many hearts are covered up?"

6. **Show Me.** Ask students to count on with you. Say, "Let's keep on counting from where Johnna left off. You help me. Who can come and point while we count on from 7 together? What number comes next?"

7. Have at least one student come and point while the class counts on.

8. If students are uncertain of what number should come next, say, "I'll do Johnna's part, and then you can count the rest."

9. Count from 1 through 6 slowly without pointing. Then point to the 7 in the image and emphasize the word by stretching it out saying "s – e – v – e – n." Gesture to students to indicate that they should point and count the rest of the hearts from there.

Note: Consider using the Show Me tool (see Appendix B).

FACILITATE

1. Tell students they get to practice counting the way Johnna and Terrence were counting. They will work with a partner.

2. Model the partner roles with a student using one of the Counting Cards.

   » The pair will begin with the folded card (see Figure 4.3). The number will tell how many things have already been counted.

Figure 4.3 Folded Counting Card

» Partner 1 will count the rest of the hearts, counting on from that number (8, 9, 10, 11).

» Partner 1 will say how many hearts there are ("There are 11 hearts!").

» Partner 2 will unfold the card and count all to prove that there are that many (see Figure 4.4).

Figure 4.4 Unfolded Counting Card

» Partner 2 will say if they agree or disagree ("I agree! There are 11 hearts!").

» Then the pair can get a new card and trade roles.

3. **Observe/Interview.** Circulate from pair to pair noting how the students approach the task. Are they confident with counting on from the starting number, or are they orally recounting the initial set from zero before they count on the additional figures?

! PRODUCTIVE STRUGGLE

Allow students to count all if needed, but encourage them to *also* try counting on from the given number. Through repeated opportunities to count on and prove that the results will be the same whether they count all or count on, students will begin to gain confidence with counting on. Consider posing a prompt such as, "I wonder what will happen if we count them all and count on?"

Ask:

» How did you know what number to say next?

» Do you have to count this one (indicate the left-most image in the "count on" group) first? Or could you start with this one (indicate a different image in that set)?

Note: Consider using the Observation and Interview tools (see Appendix B).

CLOSE: MAKE THE MATH VISIBLE

1. After each pair has had enough time to trade roles at least twice, bring the class back together.

2. Revisit the class wonderings recorded on the board during the Launch phase of the task. Ask, "Which of our wonderings were answered?"

3. Have students *Turn and Learn* from a partner. Tell them to ask their partner: "What shapes did you count? How many were there?"

4. Ask, "How did you know what number to say next when you were counting on?"

5. **Hinge Question.** Show students a paper bag with 10 items in it, but don't let them see the things inside the bag. Also show them 5 more items. Tell students: "There are 10 (blocks, counters, etc.) in this bag. I'm going to add these (indicate the 5 additional items) too. Write on your paper or marker board how you would count the rest."

TASK 1: QUICK COUNTS COUNTING CARDS

 To download printable resources for this task, visit **resources.corwin.com/ ClassroomReadyMath/K-1**

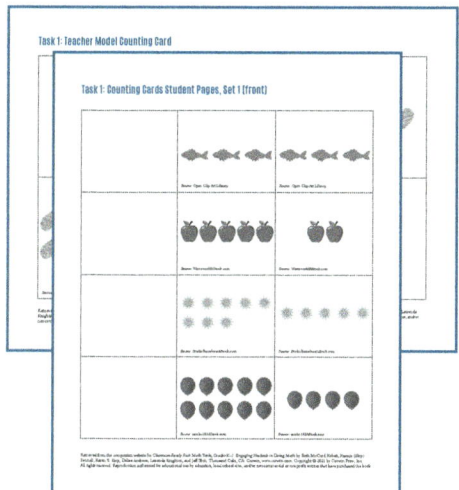

POST-TASK NOTES: REFLECTION & NEXT STEPS

Mathematics Standard

- Count to 100 by ones and by tens.

Mathematical Practices

- Construct viable arguments and critique the reasoning of others.
- Look for and make use of structure.

Vocabulary

- one/ones
- ten/tens

Materials

- connecting cubes in groups of 10 and singles or base ten materials

Race to 100

Relate counting by tens to counting by ones

TASK

Race to 100

Darla and Julio were counting together. Darla said, "Let's count to 100!" She started counting by ones using cubes (see Figure 4.5).

Figure 4.5 Darla Counting

Source: Galaxy/iStock.com

Julio said, "I think I know a faster way we can count to 100!" He started counting by tens using 10 sticks (see Figure 4.6).

Figure 4.6 Julio Counting

Source: Galaxy/iStock.com

What was different about the way Darla and Julio were counting? Is Julio's way faster? Will they still land on 100 when they are counting?

TASK PREPARATION

- Use this task after students have learned that 10 is one group of 10 ones.
- Consider how to organize students into heterogeneous groups of two or three.
- Prepare enough collections of connecting cubes (about 105 loose and 12 premade stacks of 10) for each group.

LAUNCH

1. Show students a single connecting cube and a stack of 10. Ask, "What does each represent? How are they alike? How are they different?"

2. Elicit from students that these are tools we use to represent numbers. The single cube represents "one" and the stack of 10 represents a group of 10 ones or 1 ten.

3. Display the two models side by side and facilitate a See, Think, and Wonder.

4. Record student notices on the board. Highlight:

   » There are three "things" in each picture. The first picture is three ones, the second picture is three groups of 10.

   » One shows counting by ones, the other shows counting by tens.

   » The count-by-ones picture shows fewer cubes than the count-by-tens picture: 3 < 30.

FACILITATE

1. Share the full task story with students. Ask, "Which way to count do you think is faster? Will both ways still help us count 100 in all?"

2. Organize students into pairs or small groups.

3. Provide each group with a collection of cubes and ask them to test out both methods (Darla's way and Julio's way) for counting to 100.

4. **Show Me/Interview.** As you visit each group, ask:

   » "Show me 10 (10 ones or 1 ten). Can you show me another way to represent 10?"

   » "Show me 20 (20 ones or 2 tens). Can you show me another way to represent 20?"

   » "Which way would you prefer to represent 30? Why?"

PRODUCTIVE STRUGGLE

Students may tire of counting and recounting. Celebrate their perseverance when they count and recount to ensure accuracy.

Note: Consider using the Show Me and Interview tools (see Appendix B).

CLOSE: MAKE THE MATH VISIBLE

1. When students have counted both ways, bring the class back together. Using information gathered during the Show Me/Interview, ask students to share evidence of their own thinking. Highlight the following ideas:

   » It was faster to count by tens because you can count 10 ones at a time.

   » We could know we can count 100 both ways because we can also count each of the ones that make up every 10.

   » We know we can count 100 both ways because we can line them up side by side and match all 100 ones to the 100 ones that make up the 10 tens.

STRENGTHS SPOTTING

Students may share that they feel more confident counting by ones because they know for sure they have counted all of the items. Celebrate the strength in this thinking—it is a good thing to want to be confident in one's counting of a collection. Leverage this thinking to ask: "How can we start to feel confident that counting by tens is still counting everything?"

POST-TASK NOTES: REFLECTION & NEXT STEPS

Task 3

It's a Match!

Matching number names to representations

TASK

It's a Match!

Maurice and Celia were playing a matching game. Can you figure out which cards are matches? Explain how you know.

TASK PREPARATION

- Use this task as an opportunity to review the written numerals for 0–20.

- Plan for heterogeneous groups of three students.

- Organize sets of the Matching Cards for each group.

LAUNCH

1. Using one set of Matching Cards, distribute one card to each student in the class. Facilitate a *Turn and Learn*. Ask, "Turn and talk to a partner. Tell them what you know about the card you have."

2. Have several students share what they notice from looking at the cards. Elicit the following from students:

   » Some of the cards have pictures of base ten blocks.

   » Some of the cards have pictures of ten frames with different amounts of dots.

   » Some of the cards have numerals.

3. **Hinge Question.** How can we figure out what amount is represented on this card (see Figure 4.7)?

Figure 4.7 Card With Blocks

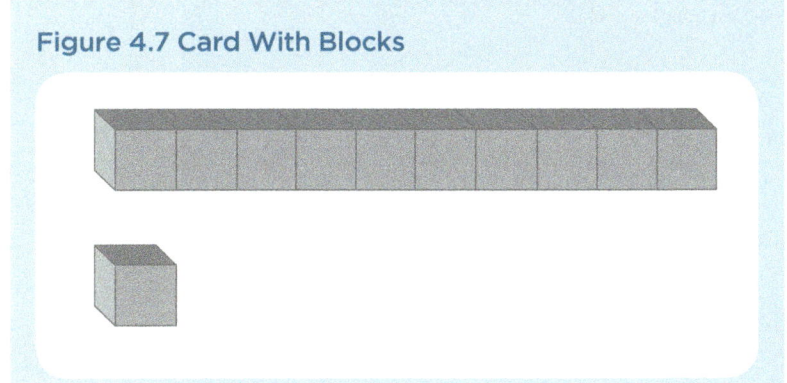

Mathematics Standard

- Write numbers from 0 to 20. Represent a number of objects with a written numeral 0–20 (with 0 representing a count of no objects).

Mathematical Practices

- Attend to precision.
- Look for and make use of structure.

Vocabulary

- base ten blocks
- ten frames

Materials

- Matching Cards student pages
- base ten blocks
- ten frames

FACILITATE

1. Present the full task to students. Ask, "Of the cards you have, which ones might be matches?"

2. Organize students into heterogeneous groups of three. Give each group a Matching Cards set.

3. All cards should be arranged *face up* where all three students can clearly see all the cards.

4. Students should take turns being the first to pick a card.

5. As each student picks any card, they should tell their group the amount their card represents (e.g., "This card shows 11." or "There are 11 dots."). Each of the other group members then tries to find another representation of the selected amount.

6. Once all three agree that the cards match, the next student begins their turn.

7. **Observe/Interview.** As students work, take note of the strategies they are using to find matches. Listen to how they read the numbers. Ask:

   » How do you know that these cards match?

   » How can you prove to me that this card shows 11?

   » What do you notice about the cards that you've matched together?

   » How is *this* 13 (indicate card with base ten materials) like *this* 13 (indicate card with ten frame)?

Note: Consider using the Observation and Interview (small group) tools for monitoring and recording student responses (see Appendix B).

CLOSE: MAKE THE MATH VISIBLE

1. Bring the class back together when all groups have matched most of the sets.

2. Based on observations/interviews, ask students to share what they notice from the activity.

3. Display a collection of matched cards (see Figure 4.8).

Figure 4.8 Matched Cards

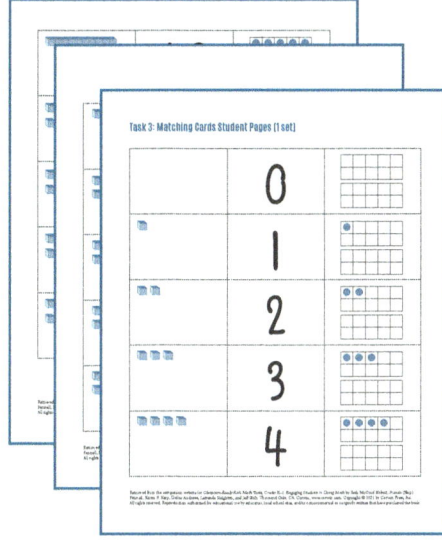

Ask

» How can we know that these cards all represent the same amount?

» How is this representation (indicate the base ten card) the same as this representation (indicate the ten frame)? How are they different?

» **Hinge Question.** If I put *one more* dot on this ten frame, what number would that represent?

TASK 3: MATCHING CARDS

 To download printable resources for this task, visit **resources.corwin.com/ ClassroomReadyMath/K-1**

POST-TASK NOTES: REFLECTION & NEXT STEPS

Mathematics Standard

- Write numbers from 0 to 20. Represent a number of objects with a written numeral 0–20 (with 0 representing a count of no objects).

Mathematical Practice

- Construct viable arguments and critique the reasoning of others.

Vocabulary

- number names 1–20
- first
- next
- before
- after

Materials

- What's in a Name Image Collection student pages (sample images)
- Photo Frames student page for student-made photos
- tape or glue sticks
- large paper (chart paper or 11 × 14/11 × 17 sheets) for each student
- markers

Task 4
What's in a Day?

Write numbers from 1 to 20

TASK

What's in a Day?

Figure 4.9 Ruben's Day

Image sources: Toothbrush by pixabay.com; School bus by luplupme/iStock.com; Kid sleeping by Rudzhan Nagiev/iStock.com

Ruben's grandma asked him what he did yesterday. He realized he was busy! He decided to make a list of all the things he did in order (see Figure 4.9). What would the list of your day look like? What things did you do yesterday? How can you use mathematics to show the order of what you did during the day?

TASK PREPARATION

- Use this task to provide opportunities for students to practice writing numbers to label events in sequence.

- Consider how you will partner students to share their thinking (see Facilitate, step 6).

- Print and cut apart collections of images for each student. Sample images are provided. Another option is to take pictures of your students doing routine daily classes and events at school (e.g., breakfast, attendance, math class, reading group, art class, PE, lunch). Print and cut apart these images along with a few taken outside of school to create a collection of images that are meaningful to students.

- If possible, collect photos to chronicle a day in your own life (see Launch, step 1).

ACCESS AND EQUITY

Consider enlisting students and their families in this project as well. Ask students to describe some things they do each day outside of school. Students might describe things like visiting a community cultural center or playing at a local park. Gather images of these kinds of events as well so students will be able to see themselves in the task.

LAUNCH

1. Bring the class together and display an unordered collection of photos (either images from the sample set or preferably images chronicling a day in your own life).

2. Facilitate a See, Think, and Wonder. Encourage students to *Turn and Learn* what a partner is thinking.

3. Let the class share the ideas they discussed with their partners. They might begin to guess about what individual pictures show. Ask, "When do you think these pictures might have been taken?"

4. As soon as students begin to share ideas connected to daily events, tell them the task story (if sharing pictures from your own day, insert yourself into the story in place of Ruben).

5. **Hinge Question.** "Which of these pictures do you think might be the very first picture I (or Ruben) took that day? Why?"

6. After students share their thinking, indicate the first picture and tell students, "I'm going to label this picture '1' because it's activity number 1 that I (or Ruben) did that day."

7. Select a couple more images and label them "2" and "3."

FACILITATE

1. Ask students to *Turn and Talk* with a partner about things that might be pictured in their day.

2. Give each student a collection of images, large paper, glue or tape, and a marker.

3. Tell students they get to choose images that represent their day and put them in order. They don't have to use all the pictures, and if they have time, they can draw others.

4. As they select each picture, they should arrange it and, when they are sure, stick it on their paper in order and label it with a number.

5. As students work, circulate through the class to

   » **Observe:**

      » How are students writing their numerals?

      » Are you noticing reversals that seem common in the group?

      » Are you noticing that several students are struggling with a particular numeral?

   » **Interview:**

      » What was picture number 3 for you?

      » Do you do _____ every morning before school?

      » What's something you do that isn't pictured here?

> ## ! PRODUCTIVE STRUGGLE
>
> The primary goal of this task is for students to practice writing the numbers 1–20, so that should be where the productive struggle exists. If students accidentally write a number more than once or skip a number, teachers may wish to simply prompt the student with the missing/correct number.

» **Show Me:**

  » Show me how you wrote this number 5.

  » Show me how you will write the next number.

  » Show me where picture number 8 is on your paper.

Note: Consider using the Observation, Interview, and Show Me tools to help organize responses and possible next steps (see Appendix B).

6. When most students have nearly 20 pictures labeled on their paper, have students meet to talk about their day with a partner.

CLOSE: MAKE THE MATH VISIBLE

1. Facilitate a Notice and Wonder Gallery Walk so students can see everyone's day represented.

2. Have students share their notices with the class.

3. Ask specific questions of the class to highlight and practice any numbers that need review. For example:

  » What do you notice about picture number 6 on everyone's page. Did everyone have the same number 6 picture? Why?

  » Number 6 looks a lot like the number 9, doesn't it? Sometimes it's tricky to remember which direction to write it. How do we write the number 6? Let's practice!

TASK 4: WHAT'S IN A DAY IMAGES AND PHOTO FRAMES

 To download printable resources for this task, visit **resources.corwin.com/ ClassroomReadyMath/K-1**

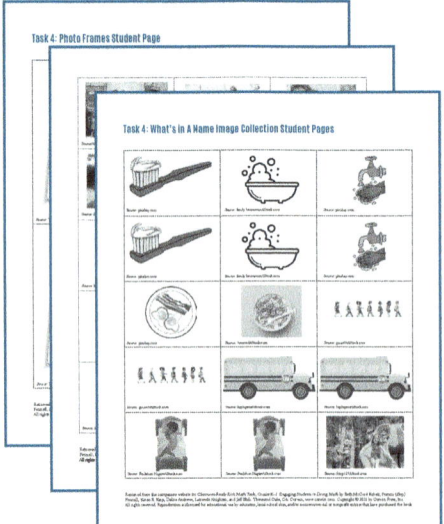

POST-TASK NOTES: REFLECTION & NEXT STEPS

Counting and Cardinality

Counting Objects

TASK 5: KINDERGARTEN: MICE ON THE MOVE

Understand the relationship between numbers and quantities; connect counting to cardinality.

a. When counting objects, say the number names in the standard order, pairing each object with one and only one number name and each number name with one and only one object.

b. Understand that the last number name said tells the number of objects counted. The number of objects is the same regardless of their arrangement or the order in which they were counted.

c. Understand that each successive number name refers to a quantity that is one larger.

TASK 6: KINDERGARTEN: THE MAGIC WAND

Understand the relationship between numbers and quantities; connect counting to cardinality.

a. When counting objects, say the number names in the standard order, pairing each object with one and only one number name and each number name with one and only one object.

b. Understand that the last number name said tells the number of objects counted. The number of objects is the same regardless of their arrangement or the order in which they were counted.

c. Understand that each successive number name refers to a quantity that is one larger.

TASK 7: KINDERGARTEN: COLOR COUNTING

Understand the relationship between numbers and quantities; connect counting to cardinality.

a. When counting objects, say the number names in the standard order, pairing each object with one and only one number name and each number name with one and only one object.

b. Understand that the last number name said tells the number of objects counted. The number of objects is the same regardless of their arrangement or the order in which they were counted.

Anticipating Student Thinking: For most young learners, mathematics begins with counting. Kindergarten mathematics experiences help in developing understandings related to the counting process. The tasks in this chapter connect counting to cardinality, the number of objects counted. All three of the tasks in this chapter engage students in using manipulative materials in connecting oral counting to counting objects. And, each task is truly an enjoyable activity! Use Observation to monitor the progress of all of your students in the chapter's tasks. The use of Interview and Show Me also provides opportunities for less vocal students to talk about and demonstrate their understanding of the counting process. These engaging tasks provide you with an early benchmark of the mathematical strengths of your students.

THINK ABOUT IT

As you consider adapting the chapter tasks to meet the needs of your students, you may consider extending the number of objects to be counted to as many as 10 objects. A related adaptation may be to have students start their counting from a given amount to a designated number of counters (e.g., start with three objects and count to five objects).

Mathematics Standard(s)

- Understand the relationship between numbers and quantities; connect counting to cardinality.

  a. When counting objects, say the number names in the standard order, pairing each object with one and only one number name and each number name with one and only one object.

  b. Understand that the last number name said tells the number of objects counted. The number of objects is the same regardless of their arrangement or the order in which they were counted.

  c. Understand that each successive number name refers to a quantity that is one larger.

Mathematical Practice

- Model with mathematics.

Vocabulary

- count
- order

Materials

- Mice Counter Cards student page, one set per pair
- counting cubes
- paper cups
- dice (optional)
- *Mouse Count* by Ellen Stoll Walsh could also be used with this task

Task 5
Mice on the Move

Count to 5

TASK

Mice on the Move

Figure 5.1 Mice Playing Outside

Source: Tree by belander/iStock.com; House by rambo182/iStock.com; Mice by mariaflaya/iStock.com

Five mice are playing outside by the tree (see Figure 5.1). Their mother calls them to come home. Using your counting cubes, count the mice as they walk from the tree to their home.

TASK PREPARATION

- Ensure that students have at least five counting cubes each as well as a paper cup to represent the mouse house.

LAUNCH

1. Bring students together as a whole group and give each of them a cup large enough to hold five or more connecting cubes.

> **ALTERNATE LEARNING ENVIRONMENT**
>
> Students could use virtual counting cubes to provide access to manipulatives in a remote learning setting: www.didax.com/apps/unifix/

2. Tell students, "Today we are going to act out a story about mice. I want you to help your mice go home. We are going to use a cup to be their home."

3. Say, "Let's all help two mice go home." Have students pick up one cube and put in the cup and model the count "One." Let them pick up another cube and put it in the cup and count "Two! We helped these two mice go home!"

FACILITATE

1. Place students in pairs and distribute a collection of Mice Counter Cards or cubes to use as counters.

2. Say, "You and your partner will work together to solve this task: Five mice are playing outside by the tree. Their mother calls them to come home. Using your counting cubes, count the mice as they walk from the tree to their home."

3. Observe students to see how they count (one-to-one counting, rote counting).

4. Adjust the amount for students by giving them a new number or by having them roll a die to determine the number of mice that need to go home.

CLOSE: MAKE THE MATH VISIBLE

1. Bring students back together for a group discussion.

2. **Show Me.** Have the students demonstrate their counting. Ask, "How do you know when you have 5 in the cup?"

3. Show an incorrect count by double-counting one mouse and ask, "Is that the right way to count the mice? Why or why not?"

4. Emphasize their correct counting steps (pointing, one-to-one, moving items while counting).

5. **Hinge Question.** Show the students that you have 3 counters in a cup. Ask, "If I put another counter in the cup, how many are in the cup? Is that more or less than before? How many more?"

Note: Consider using the Show Me tool (see Appendix B).

> ### STRENGTHS SPOTTING
>
> When students notice a mistake and note that mistake by adjusting a strategy, they are demonstrating strategic thinking. In addition, when working through an incorrect count, students can recognize that making mistakes is part of learning, a dispositional strength. Take the time to celebrate and recognize these strengths.

TASK 5: MICE COUNTER CARDS STUDENT PAGE

 To download printable resources for this task, visit **resources.corwin.com/ ClassroomReadyMath/K-1**

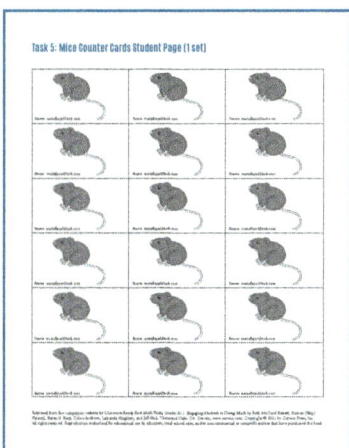

POST-TASK NOTES: REFLECTION & NEXT STEPS

Mathematics Standard(s)

- Understand the relationship between numbers and quantities; connect counting to cardinality.

  a. When counting objects, say the number names in the standard order, pairing each object with one and only one number name and each number name with one and only one object.

  b. Understand that the last number name said tells the number of objects counted. The number of objects is the same regardless of their arrangement or the order in which they were counted.

  c. Understand that each successive number name refers to a quantity that is one larger.

Mathematical Practice

- Model with mathematics.

Vocabulary

- count
- order

Materials

- counting cubes or two-color counters
- unsharpened pencils
- Magic Counting Wands student page
- cellophane tape (consider taping the wands to unsharpened pencils before the lesson)

Task 6
The Magic Wand

Count to 20

TASK

The Magic Wand

Figure 5.2 A Flower

Source: Flower by freestockphotos.biz

You have been given a magic wand that makes flowers change color! (see Figure 5.2)! Your partner will give you some flowers. Count how many flowers change color by tapping on them with your magic wand!

TASK PREPARATION

- Ensure that each pair of students has a handful of counting cubes or two-color counters as well as an unsharpened pencil to serve as the magic wand.

LAUNCH

1. Bring students together as a whole group and give each of them an unsharpened pencil. Say, "Today we are going to be wizards with a magical power. I am going to give each of you a magic wand that you can use to pretend you are changing the color of flowers. To change the color, you also must say the number as you count the flowers."

2. Gather a handful (1–9) of cubes/counters and put them in front of you. Say, "Watch me use my magic wand to change the colors of the flowers in front of me as I count." As you tap each cube with the pencil (wand), count out loud until you have counted all of the cubes. If using the two-color counters, flip over the counter with each tap of the wand.

FACILITATE

1. Place students in pairs and distribute cubes/two-color counters.

2. Say, "You and your partner will each take turns with your magic wands." Have one person gather a handful of flowers (1–20). The other person will use their magic wand to change the colors and count the flowers. When one person finishes, switch roles.

3. **Observe** student pairs as they count, asking selected pairs to **Show Me** how they are changing the flowers' colors.

Note: Consider using the Observation tool (see Appendix B).

ALTERNATE LEARNING ENVIRONMENT

Students could use the two-color counter virtual manipulative to engage in this activity in a virtual setting: www .didax.com/apps/two-color-counters/

CLOSE: MAKE THE MATH VISIBLE

1. Bring students back together for a whole-group discussion. Allow students to demonstrate their counting.

2. Show an incorrect count by double-counting a counter or skipping a number. Ask the students to *Turn and Talk*. Ask, "What did you notice?"

3. Ask, "Is that the right way to count? Why or why not? Turn and talk with a partner about how I should count."

4. Elicit from the students the correct counting steps by emphasizing how the wand touches each cube.

5. **Hinge Question.** Show a collection of flowers to the students and ask them to orally count as you point to the counters. After they count to a number, ask, "What will the next number be? Is it larger or smaller than the number we just counted?"

ACCESS AND EQUITY

Partner talk before whole-class sharing helps to ensure that every student has an opportunity to think before the whole-group discussion and gives the teacher an opportunity to circulate and listen for two or three ideas to highlight with the whole group. If the teacher wants particular students to share, give them time to rehearse before sharing in the whole-group settings. Students should be allowed to share their own ideas or an idea they heard from their partner.

TASK 6: MAGIC COUNTING WAND STUDENT PAGE

 To download printable resources for this task, visit **resources.corwin.com/ ClassroomReadyMath/K-1**

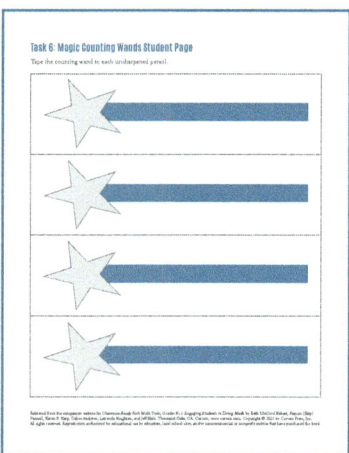

POST-TASK NOTES: REFLECTION & NEXT STEPS

Task 7
Color Counting

Count start and end points; object arrangement and quantity

TASK

Color Counting

Figure 5.3 A Bag of Blocks

Source: Joe_Potato/iStock.com

Gil and Marilyn have a bag of blocks (See Figure 5.3). Gil counts the blocks starting with a red block. He says, "There are 5 blocks!" Marilyn says, "I wonder if it will be different if we start with a blue block instead." What do you think? Will the number of blocks change?

TASK PREPARATION

- Prepare baggies of buttons or blocks for each student so that each button/block is a different color. The number can be adjusted for each student.

LAUNCH

1. Bring students together as a whole group. Tell students, "Today I have a story about counting blocks!"

2. Engage the students in some questions. Ask, "What does counting tell us?" (how many).

3. Say, "Gil and Marilyn have a bag of blocks. Gil counts the blocks starting with a red block. He says, 'There are 5 blocks!' Marilyn says, 'I wonder if it will be different if we start with a blue block instead.' What do you think? Will the number of blocks change? Turn and talk with a partner and discuss your idea."

Mathematics Standard(s)

- Understand the relationship between numbers and quantities; connect counting to cardinality.

  a. When counting objects, say the number names in the standard order, pairing each object with one and only one number name and each number name with one and only one object.

  b. Understand that the last number name said tells the number of objects counted. The number of objects is the same regardless of their arrangement or the order in which they were counted.

Mathematical Practice

- Model with mathematics.

Vocabulary

- count

Materials

- baggies of different-colored buttons or blocks (each block must be a different color)

- Shape Outlines student pages

FACILITATE

1. Say, "I am going to give each of you some blocks that are all different colors. Take your blocks out of the baggies and put them in front of you."

2. Say, "Let's start with the red block and count how many blocks we have. Write down that number."

3. Say, "This time, let's start with the blue block and count how many blocks we have. What do you notice? How many did you count? Is it the same number as before?"

4. Ask, "What will happen if we start our count from a different color? Repeat with a different color."

5. **Observe** students to see how they count.

6. Next, bring the students back to a whole-class discussion. Ask, "Will counting our blocks when they are arranged in a shape change how many there are? Turn and talk with a partner to share your idea."

7. Elicit predictions from the students.

8. Say, "I am going to give each of you a square and a circle shape cards. Place your blocks on the line that make the shape." Model how to do this if needed (see Figure 5.4).

> ## ! PRODUCTIVE STRUGGLE
>
> Asking students to pose ideas and predictions helps them understand that predicting and testing ideas is part of developing mathematical thinking.

Figure 5.4 Sample Shapes

9. Say, "Let's start on the corner of the square and count. How many did you count?" Next, say, "Put the same counters that you put on the outline of the square on the line that makes the circle. Let's count our blocks starting at the top of the circle." Ask, "How many did you count? What do you notice?"

Note: Consider using the Observation tool (see Appendix B).

CLOSE: MAKE THE MATH VISIBLE

1. Bring students back together for a whole-group discussion. Allow students to demonstrate their counting. Emphasize their correct counting steps by asking the following questions:

   » How many blocks did we have when we started counting with the red one?

   » How many blocks did we have when we started counting with the blue one?

   » Did you think the answers would be the same? Why or why not?

2. **Hinge Question.** Does it matter what color block we start with, when we count all of the blocks?

3. Show an incorrect count by double-counting one block and ask, "Is that the right way to count the blocks? Why or why not?"

4. Emphasize their correct counting steps (pointing, one-to-one, moving items while counting).

PRODUCTIVE STRUGGLE

Support sense making by encouraging students to consider counterexamples, and ask questions to focus students on the correct and incorrect counting methods.

TASK 7: SHAPE OUTLINE STUDENT PAGES

 To download printable resources for this task, visit **resources.corwin.com/ ClassroomReadyMath/K-1**

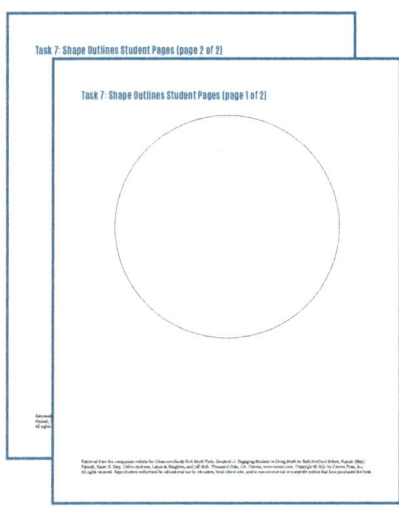

POST-TASK NOTES: REFLECTION & NEXT STEPS

Operations and Algebraic Thinking

Patterns and Relationships

TASK 8: KINDERGARTEN: EGGS-PLOSION

Count to answer "how many?" questions about as many as 20 things arranged in a line, a rectangular array, or a circle, or as many as 10 things in a scattered configuration; given a number from 1–20, count out that many objects.

TASK 9: KINDERGARTEN: SHAKE IT UP

Understand the relationship between numbers and quantities; connect counting to cardinality.

a. Understand that the last number name said tells the number of objects counted. The number of objects is the same regardless of their arrangement or the order in which they were counted.

TASK 10: GRADE 1: HAVE SOME MORE

Count to 120, starting at any number less than 120. In this range, read and write numerals and represent a number of objects with a written numeral.

Anticipating Student Thinking: The three tasks provided in this chapter continue student recognition of and emphasis on the foundational importance of counting. Tasks 8 and 9 focus on kindergarten-level standards related to counting. These tasks engage students in counting varied configurations of eggs (Task 8), and in understanding that the last number name stated when counting, in this case, collections of objects, indicates the number of objects counted (Task 9). Task 10 is a first-grade task that will engage students in using base ten blocks to both represent and count numbers that begin with numbers less than 100 and extend beyond this important numerical benchmark. As you anticipate student engagement in these tasks, consider the importance of how numbers in each of these tasks will be represented.

THINK ABOUT IT

As you plan for and implement the Launch, Facilitate, and Close of the chapter's task lessons, consider the student struggles that may occur as they are engaged in a task. Such pre-task considerations should serve as a reminder to you as you prepare to support your students in each of these counting-related tasks.

Task 8
Eggs-plosion

Count physical objects in various arrangements

TASK

Eggs-plosion!

Figure 6.1 One of Ms. Dunham's Chickens

Source: Marcos Assis/iStock.com

Ms. Dunham is raising chickens (see Figure 6.1). Last week, there were a lot of eggs! There are so many eggs that she didn't have places to put them all. Ms. Dunham is going to make egg salad to take to the community center. She needs to count all the eggs so she knows how much of the other ingredients to use. We are going to help Ms. Dunham count the eggs!

TASK PREPARATION

- Plan for heterogeneous student pairs.

- Raid your kitchen and storage spaces to find a collection of assorted containers for the activity. Look for tube pans (e.g., Bundt®, angel food); cupcake pans; large and small cake pans; and longer, more narrow trays (such as a pencil or cookie tray). Craft/grocery stores also sell cardboard or foil cake pans if preferred.

- Visit your local grocer or farmers market to collect a variety of egg cartons (e.g., 6, 12, and 18 ct.)

- Organize various amounts of eggs into each container. Egg cartons and cupcake pans will create arrays, tube pans will create circles, pencil trays will create rows, and other pans will create scattered collections.

LAUNCH

1. Bring the class together and show them the collections of eggs (see Figure 6.2). Facilitate a Notice and Wonder.

Figure 6.2 Eggs

Source: Eggs in circle by Ninell_Art/iStock.com; Eggs in carton by MEDITERRANEAN/iStock.com; Eggs in tray by DmitriyKazitsyn/iStock .com; Eggs in row by Bozena_Fulawka/iStock.com; Randomly placed eggs by bjdlzx/iStock.com; Bundt pan by ninikas/iStock.com; Tray by vikif/iStock.com

2. Show students the image and ask students what hens do (lay eggs). Ask, "Have you ever eaten eggs?" Encourage students to describe some of the different ways they eat eggs.

3. Tell students the story from the task.

4. Ask, "How many eggs do you think there might be?" Encourage students to explain how they made their estimates.

5. Record student estimates on the board.

FACILITATE

1. Ask, "How could we count all of these eggs?"

2. Organize students into heterogeneous pairs. Give each pair at least two different types of containers of eggs to count.

3. If you notice students are finding it challenging to keep track of which eggs are counted, consider offering the option of a tool (dry-erase marker, dot sticker, sticky note, etc.) to mark eggs as they are counted.

4. **Observe.** As students count eggs, take note of their strategies. Are they

   » moving the eggs to a new location as they count (taking them out of or putting them into the container)?

   » keeping track of where they started counting (especially with the tube pans/circles of eggs)?

   » counting in sets (e.g., skip counting by 2s or 5s, etc.)?

   » concerned about color or orientation of each egg?

> ! **PRODUCTIVE STRUGGLE**
>
> Keep this support option (Facilitate, step 3) ready to go, but do not set up the task with this protocol in place. Allow students to spend a bit of time working on their own to figure out how to manage counting accurately. Even if support becomes needed, ask students if the support would be helpful, offer it on an individual basis, and let students manage the process.

5. **Interview.** Ask,

» How many eggs are in this container (point to a specific container)?

» *Record counts for each pair as you go.*

» How did you decide where to start counting?

» Did you count them in a specific order? Why?

» Which container was easier to count? Why?

ACCESS AND EQUITY

The focus of this task is the act of counting out the sets of eggs, not writing or recognizing numbers. Eliminate a potential barrier to task access by recording the counts students provide to you orally.

6. *Pair-to-Pair Share:* As groups finish counting, have them meet up with another pair to compare their containers and counts.

Note: Consider using the Observation and Interview (small group) tools for monitoring and recording student responses (see Appendix B).

CLOSE: MAKE THE MATH VISIBLE

1. When all the egg collections have been counted and counts recorded by the teacher, bring the class back together.

2. Based on observations and interviews, ask student pairs to share which kinds of collections were easiest for them to count and why. Elicit the following ideas from students:

» It's important to keep track of which eggs are already counted so that you don't count the same egg more than once.

» It's important to keep track of which eggs haven't been counted yet so that you don't miss any eggs.

» The arrays (cartons/cupcake tins) might be easier to count and keep track because the organization makes it obvious which eggs are or are not counted.

Revisit the predictions the class made for the total number of eggs. Ask, "Do these predictions still seem reasonable? Does anyone want to adjust their prediction?"

STRENGTHS SPOTTING

Students with strengths in adaptive reasoning (NRC, 2001) will demonstrate comfort with reflecting on the success of their approach and enjoy the opportunity to describe how they shifted strategies while solving the task.

3. Show students the list of counts they provided. Display your calculator and compute the total using the pairs' counts.

4. Compare the results to the students' predictions and allow students to discuss the results. Ask, "Were our predictions pretty close? Were there more eggs or fewer eggs than we thought there would be?"

POST-TASK NOTES: REFLECTION & NEXT STEPS

Kindergarten

Mathematics Standard(s)

- Understand the relationship between numbers and quantities; connect counting to cardinality.

  a. Understand that the last number name said tells the number of objects counted. The number of objects is the same regardless of their arrangement or the order in which they were counted.

Mathematical Practice

- Attend to precision.

Vocabulary

- similar
- same amount

Materials

- collections of small objects in various sizes and colors (e.g., Legos®, blocks, stickers, pencils)
- paper bags

Task 9
Shake It Up

Understand that the number of objects counted stays the same

TASK

Shake It Up

Figure 6.3 Ana's Lego® Collection

Source: Paper bag by chengyuzheng/iStock.com; Blocks by KariHoglund/iStock.com.

Ana counted her Lego® collection. After she counted it, her dad placed the Legos® into a bag (see Figure 6.3). If Ana pours them out of the bag, will she still have the same number of Legos®?

TASK PREPARATION

- Consider how you will organize students into heterogeneous pairs.

- Prepare a variety of collections of up to 20 objects in small paper bags. Make enough for each pair to have their own collection. Vary the total number of objects in each collection.

- Prepare a bag with a collection for the class to count together (see Launch, step 1). You might want this to be a larger bag with larger items to ensure that all students are able to see the objects and count together.

ACCESS AND EQUITY

If there are no Legos® in your classroom, edit the story to include some other small countable object familiar to your students. Be sure to use something that comes in a variety of sizes and colors such as snap cubes.

LAUNCH

1. Show the class a bag containing a collection of similar items. Consider seating the class in a circle so the collection can be displayed in the center and all students can see easily.

2. Pour out the collection onto the floor and ask students to count it with you. Model effective counting practices by touching and moving each item as you count it.

3. Write the total number of items on the board.

4. Present the task story to the class and ask, "Do you think she will still have the same number of Legos®?"

5. Have students *Turn and Talk* with a partner.

6. Collect the items the class counted and put them back in the bag. Shake it up and pour it back out onto the floor.

7. Ask, "I wonder if we still have the same amount here?"

8. Tell students you will check on that later, but first they will get to test out their ideas by counting their own collections. *(You may wish to put the objects back in the bag to limit distractions during the Facilitate phase of the task.)*

FACILITATE

1. Organize the class into heterogeneous pairs and distribute a collection of objects to each pair.

2. Direct students to work together to

   » count their collection,

   » write the number down,

   » put the collection back in their bag and shake it up, and

   » pour the collection back out and recount to see if they still have the same number.

> ### ACCESS AND EQUITY
> Consider providing printed or Braille number cards for students with vision or dexterity needs.

3. Students should shake and recount their collection at least three to four times until they are convinced there will always be the same amount.

4. As students work, circulate to **Interview/Show Me:**

   » How many items did you count? Show me how you know there are that many.

   » How many items do you think there will be after you shake up the bag? Why?

   » When you pour them out of the bag, they look different—that red one was over here before. Does that mean there are a different number now? Why or why not?

Note: Consider using the Interview (small group) and Show Me tools for monitoring and recording student responses (see Appendix B).

CLOSE: MAKE THE MATH VISIBLE

1. After the class is generally convinced that shaking up their collection will not change the total number they have, bring students back together on the carpet.

2. Select students to share their thinking based on the interviews conducted during the Facilitate phase.

STRENGTHS SPOTTING

Recognize students' communication strengths when they support their ideas with mathematical details. Some students can demonstrate this strength by explaining others' ideas and asking explicit questions.

3. Say to the class, "So let's go back to our original collection. When we counted it the first time, we had (#)." *Indicate the number written on the board.*

4. Ask, "Are you telling me that no matter how I shake this up, there will still be (#) in the bag?"

5. Have students *Turn and Talk* to explain to a partner why that's true.

6. Pour out the collection and have the class count with you.

7. Celebrate their discovery!

POST-TASK NOTES: REFLECTION & NEXT STEPS

Task 10
Have Some More

Count on from 90 to 120

TASK

Have Some More!

Teri used some base ten blocks to represent the number 95 (see Figure 6.4).

Figure 6.4 Teri's Blocks

Teri's brother came by and said, "Have some more!" He gave Teri one more ten and 9 more ones (see Figure 6.5).

Figure 6.5 Blocks from Teri's Brother

How can Teri count on to find the total number represented now?

TASK PREPARATION

• Prepare a collection of base ten blocks for each student. Each collection should have 9 tens and 0–9 ones.

• Make additional base ten blocks (tens and ones) available for partners to use.

LAUNCH

1. Represent the number 86 using base ten blocks. Display the representation so that every student in the class is able to see it clearly.

Mathematics Standard

• Count to 120, starting at any number less than 120. In this range, read and write numerals and represent a number of objects with a written numeral.

Mathematical Practice

• Attend to precision.

Vocabulary

• representation/represents
• model

Materials

• base ten blocks grouped into baggies with an amount between 90–99 for each pair

• Have Some More student page, one per student

2. Have students *Turn and Talk* about what number the model represents.

3. Encourage the class to share out what number is represented.

4. Ask, "How do you know that we have represented 86?"

5. Ask, "If I add two more cubes to this set, what number would we have now?"

6. Ask a student to come up and model how to count on to prove that the new number is 88.

7. Present the task story to the class. Ask, "Could Teri count on like we did to find the new number?"

FACILITATE

1. Organize students into heterogeneous pairs.

2. Each student will get a collection of base ten blocks in a baggie representing a number from 90–99.

3. Direct students to make a math sketch and write the number in the first box on their student page.

4. Next, students take turns:

   » Student A gives Student B more base ten blocks representing a value from 9–19 and says, "Have some more!" These should come from a new location, not from either of the students' original collections. If available, these might be a different color or material to help students distinguish.

   » Student B writes down how many more they were given. Then Student B counts on from their original amount to find the new total.

   » Partners switch roles.

5. As students work together, circulate to **Observe:**

   » Are students counting by ones until they get to a decade number and then counting by tens if available?

   » Are students counting by ones all the time, even when counting a ten block?

   » Are students confident counting from 99 to 100 to 101?

   » Are students able to count by tens from a non-decade number?

> ! **PRODUCTIVE STRUGGLE**
>
> As you observe, resist the urge to intervene and make corrections. Use the opportunity to attend to student thinking with genuine curiosity and interest. Pay close attention to how students work through miscounts that occur. Look for strengths that can be leveraged to help the whole class move forward.

Note: Consider using the Observation tool for monitoring and recording student responses (see Appendix B).

6. Early finishers can get a new collection/student page and try again. Consider mixing up partners for the second round.

CLOSE: MAKE THE MATH VISIBLE

1. After each student has had at least one chance to count on, bring the class back together.

2. Based on your observations, select a few students to highlight the different counting strategies you noticed.

3. If miscounts were observed, bring those to the class's attention for discussion. For example, "I noticed that our class is really thinking about what number comes after 99. I saw so much great thinking. Some of us were counting 90 next, and some of us were counting 100 next. Turn and talk with your partner. What's a tool we've used in our classroom that could help us know the next number after 99?"

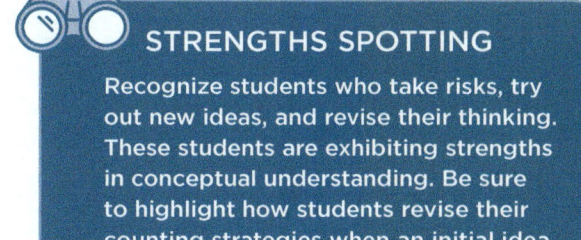

STRENGTHS SPOTTING

Recognize students who take risks, try out new ideas, and revise their thinking. These students are exhibiting strengths in conceptual understanding. Be sure to highlight how students revise their counting strategies when an initial idea doesn't work or make sense.

4. **Exit Task.** Revisit the task presented in the Launch phase (step 7). Challenge students to use a strategy that has been discussed to find the answer.

TASK 10: HAVE SOME MORE STUDENT PAGE

 To download printable resources for this task, visit **resources.corwin.com/ ClassroomReadyMath/K-1**

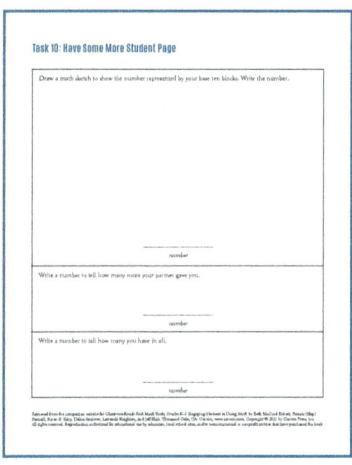

POST-TASK NOTES: REFLECTION & NEXT STEPS

Counting and Comparing

TASK 11: KINDERGARTEN: FLOWER POWER

Identify whether the number of objects in one group is greater than, less than, or equal to the number of objects in another group, e.g., by using matching and counting strategies.

TASK 12: KINDERGARTEN: GRAB, COUNT, AND COMPARE

Identify whether the number of objects in one group is greater than, less than, or equal to the number of objects in another group, e.g., by using matching and counting strategies.

TASK 13: KINDERGARTEN: THE FISHING CONTEST

Compare two numbers between 1 and 10 presented as written numerals.

Anticipating Student Thinking: This chapter extends prior experiences involving counting and cardinality to include comparing the number of objects in groups, and the comparison of numbers between 1 and 10. The chapter's three tasks, all at the kindergarten level, will engage your students in using pictured and physical representations when comparing objects in Tasks 11 and 12. In Task 13, the students will compare numerical amounts. Comparing amounts is a natural extension of and connection to counting, and many of your students have had informal experiences involving such comparisons. The chapter's tasks provide activities that introduce them to vocabulary connected to the comparison process—greater than, less than, and equal. The task lessons also suggest that students are organized in pairs as they engage the tasks.

THINK ABOUT IT

Consider all of the aspects of task preparation as you move toward implementation of the chapter's tasks. This is not limited to, but will include, student access to materials, how you will organize your class into student pairs, and the instructional time allotted to the task lesson's Launch, Facilitate, and Close. Note that if the task's implementation is virtual or online, attention to student access to, for example, materials will need to be adapted to consider resources students and families have at home or can be guided to online.

Mathematics Standard

- Identify whether the number of objects in one group is greater than, less than, or equal to the number of objects in another group, e.g., by using matching and counting strategies.

Mathematical Practices

- Model with mathematics.
- Attend to precision.

Vocabulary

- greater than, more
- less than, fewer
- same as, equal
- compare

Materials

- red and yellow counters or connecting cubes in baggies
- two paper bags with blue and green connecting cubes inside
- crayons
- Flower Power student page, one per student

Task 11
Flower Power

Compare two quantities

TASK

Flower Power

Figure 7.1 Flowers for Eliza's Teacher

Source: Kathykonkle/iStock.com

Eliza plans to give her teacher a bouquet of flowers (see Figure 7.1). She picked some red roses and some yellow roses from her garden. You have a baggie containing red and yellow cubes to represent the roses. Which color rose does Eliza have more of? Show how you know.

TASK PREPARATION

- Students will work in pairs for this task.
- Prepare two brown paper bags, one containing 8 green connecting cubes and one containing 10 blue connecting cubes.
- Place 7 red cubes and 5 yellow cubes in plastic baggies and prepare enough baggies for each pair of students.

LAUNCH

1. Bring students together as a group. Take the blue and green connecting cubes out of the two brown bags and put them into two separate piles. (Students should be seated so that everyone can see the connecting cubes.)

Conduct a Notice and Wonder routine with the students by asking students what they notice and wonder. Record their ideas on a chart:

Notice	Wonder

2. Elicit from the students that the piles of cubes have different amounts.

3. Ask, "Are there more, fewer, or the same number of blue and green cubes? Turn and talk to your partner and share your thinking."

4. Let the students suggest how they might figure out which pile has more. Ask, "How can you figure out which pile has more? Is there a way we can organize the cubes to find out? Turn and talk with your partner."

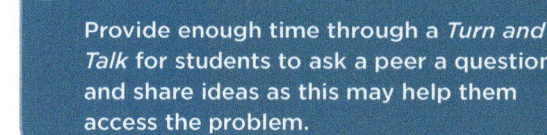

ACCESS AND EQUITY

Provide enough time through a *Turn and Talk* for students to ask a peer a question and share ideas as this may help them access the problem.

5. Solicit ideas from the students. If not suggested by a student, line up the green and blue cubes next to each other in a one-to-one correspondence. Say, "I'm going to line up the cubes so that each green cube has a blue cube as a partner."

6. Ask, "Do we have more, fewer, or the same number of blue and green cubes?" Allow time for students to share their thinking. **Show Me** how you know.

7. Ask, "Did lining up the cubes this way help you see which group has more? How? Can anyone think of a different way to find out if there are more, fewer, or the same number of blue and green cubes?" (stack into a tower and compare heights)

8. Ask, "How can we use counting to find out which group has more?"

9. Ask, "How can we use counting to find out which group has fewer?"

10. Ask, "How can we use counting to find out if groups are the same?"

FACILITATE

1. Place students in pairs and distribute baggies of red and yellow cubes and the Flower Power student page. Say, "We just solved a *compare* task together. Now, you and your partner will work together to solve a *compare* task on your own."

2. Say, "You and your partner will work together to solve this task:
Eliza plans to give her teacher a bouquet of flowers. She picked some red roses and some yellow roses from her garden. You have a baggie containing red and yellow cubes to represent the roses. Which color rose does Eliza have more of? Show how you know."

3. **Observe** student pairs to see if they use counting and/or one-to-one matching strategies to compare.

4. Place students in groups of four for a *Pair-to-Pair Share*. Tell students: "You will take turns sharing your answers to Eliza's task with the other members of your team. You and your partner will have two minutes to share your work with your team. Be sure to show how you know whether Eliza picked more red or yellow roses." As students share their

work in their groups, observe student groups to make note of the strategies students used and if students are able to show and explain their work.

Note: Consider using the Observation tool (see Appendix B).

CLOSE: MAKE THE MATH VISIBLE

1. Bring students back together as a group. Allow students to share their findings. Students should share that Eliza has more red roses than yellow roses. Allow students to show how they found their answer. Questions to ask students could include: "How do you know if there were more red or yellow roses? Did anyone use a different strategy to find the answer? Can you show the group how you found your answer? Did anyone use counting to find the answer? Did anyone match the counters one-to-one to build towers to represent the roses?" Encourage students who used one strategy to try another.

2. **Hinge Question.** How could you tell if the same number of red and yellow roses were picked?

TASK 11: FLOWER POWER STUDENT PAGE

 To download printable resources for this task, visit **resources.corwin.com/ ClassroomReadyMath/K-1**

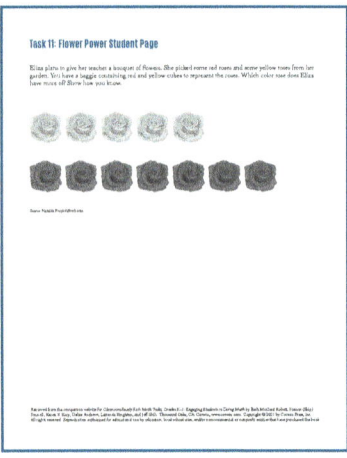

POST-TASK NOTES: REFLECTION & NEXT STEPS

_____

_____

_____

Task 12

Grab, Count, and Compare

Count and make comparisons

TASK

Grab, Count, and Compare

Jasmine and Tarik played Grab, Count, and Compare in math class. Each of them used one hand to grab a handful of counters. Now they need your help to count and compare the handfuls to see if Jasmine has the same number of, more, or fewer counters than Tarik.

Figure 7.2 Plastic Baggie of Counters

Source: Anna Pogrebkova/iStock.com

You and your partner each have a plastic baggie with the same number of counters Jasmine and Tarik had (see Figure 7.2). The red counters are Jasmine's and the blue counters are Tarik's. Compare their counters. Who has more or fewer, or do they have the same number of counters? Show how you know.

TASK PREPARATION

• Students will work in pairs for this task.

• Prepare the baggies for the partner task ahead of time.

LAUNCH

1. Bring students together as a group. Tell students, "We're going to play a new game called Grab, Count, and Compare." Students should be seated so that everyone can see. Invite a student volunteer to use one hand to grab a handful of red counters from the bag of red counters. Invite a different student volunteer to grab a handful of blue counters from the bag of blue counters.

2. Have the class talk about how they can compare the handfuls before counting. Have students share either a one-to-one matching strategy or a counting strategy to identify how many red and blue counters they have. Record the number of red counters and the number of blue counters.

Mathematics Standard

• Identify whether the number of objects in one group is greater than, less than, or equal to the number of objects in another group, e.g., by using matching and counting strategies.

Mathematical Practices

• Model with mathematics.

• Attend to precision.

Vocabulary

• compare
• greater than, more than
• less than, less
• fewer than, fewer
• same

Materials

• one brown bag or cup of 10 red counters

• one brown bag or cup of 10 blue counters

• plastic baggie of 7 red counters and plastic baggie of 9 blue counters for each pair of students

• crayons or pencils

• blank paper

3. Say, "Let's compare the red and blue counters. Are there more, fewer, or the same number of red and blue counters? Turn and talk to a partner and share your thinking. How do you know? Can anyone think of a different way to compare the counters?"

4. If no one suggests using matching, ask, "Can someone show me how to use matching to compare the counters?" Allow a student volunteer to demonstrate how to line up the counters to match them using a one-to-one correspondence. Ask, "Do we have more, fewer, or the same number of counters?" Allow time for students to share their thinking. Say, "Show me how you know."

FACILITATE

1. Place students in pairs and distribute to them baggies of red and blue counters, blank paper, and pencils or markers.

2. Say, "Jasmine and Tarik played Grab, Count, and Compare in class. Each of them used one hand to grab a handful of counters. Now they need your help to count them, and compare them to see if Jasmine has the same number of, more, or fewer counters than Tarik. You and your partner each have a plastic baggie with the same number of counters Jasmine and Tarik had. The red counters are Jasmine's and the blue counters are Tarik's."

3. Say, "Your task is to compare their counters and find out who has more or fewer, or if they have the same number of counters."

4. Observe and use the Show Me technique or conduct short Interview sessions to see if the students use counting and/or matching strategies. Are they able to quickly choose a strategy to compare the counters?

5. Say, "Use the paper and crayons to make a picture to show how you know who has more counters. Find a new partner and share your picture." Ask, "What's the same about your pictures? What's different?"

Note: Consider using the Observation, Interview, and Show Me tools (see Appendix B).

CLOSE: MAKE THE MATH VISIBLE

1. Bring students together for a whole-group discussion. Select a few student volunteers to share their work and tell how they compared. Students should share that Tarik had more counters. Ask, "What strategy did you use to find your answer? How does your picture show your work? If you used a counting strategy, try the matching strategy, and if you used a matching strategy, try a counting strategy. Did you find one strategy easier to use?"

2. Ask the whole group the following **Hinge Question:** "What if Tarik got one more counter? Would he still have more than Jasmine? Tell me how you know."

POST-TASK NOTES: REFLECTION & NEXT STEPS

Mathematics Standard

- Compare two numbers between 1 and 10 presented as written numerals.

Mathematical Practices

- Reason abstractly and quantitatively.
- Model with mathematics.

Vocabulary

- greater than, more than
- less than, fewer than
- same, equal
- number line
- number path

Materials

- Number Cards student page, one set per student
- larger number line
- Number Lines student page, one per student
- Number Path student page
- dry-erase markers
- counters
- paper plates (optional)
- sticky notes

Task 13
The Fishing Contest

Compare two numbers

TASK

Figure 7.3 The Fishing Contest

Marisha caught 5

Malia caught   8

Each said they caught the most fish.
Help them figure out who caught more.

Source: Open Clip Art Library.

TASK PREPARATION

- Students can reuse the number line templates, so prepare ahead of time by laminating the templates or placing each template in a sheet protector.

LAUNCH

1. Display the picture shown in Figure 7.4 to the students and conduct a Notice and Wonder. Record the students' ideas.

Figure 7.4 Cookies on Plates

 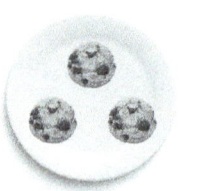

Source: Cookie by Suradech14/iStock.com; Plate by gmaydos/iStock.com

2. Elicit from the students that they are comparing the number of cookies on the two plates. Say, "We've compared groups of objects like this picture comparing cookies to find out which group has more, fewer, or the same number of objects." Ask, "Which plate has more cookies? How do you know? Which plate has fewer cookies? How do you know?"

3. Say, "Turn and talk with a partner about how you could find out which plate has more cookies and which plate has fewer cookies."

ACCESS AND EQUITY

Be sure to provide individual think time first so each student has an opportunity to make their own observations. Asking students to share their ideas with a partner before taking observations from the class gives students more time to formulate and process their ideas.

4. Allow time for students to share how they would determine which set of objects has more. If students suggest modeling with counters, allow students to demonstrate.

5. Say, "Today we are going to compare quantities using numbers only." Project or draw a similar picture to the one shown in Figure 7.5 for students.

Figure 7.5 Numbers on Plates

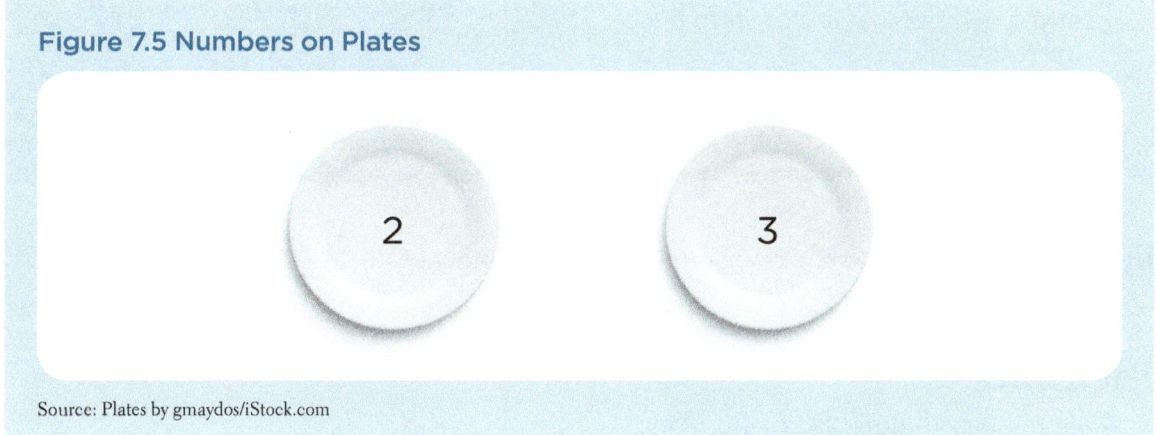

Source: Plates by gmaydos/iStock.com

6. Say, "Now we're comparing the number of cookies on the plates using numbers only. Can you think of a strategy you can use to compare when we only have numbers? Turn and talk to a partner and share your thinking." Allow time for students to share their thinking.

7. Say, "Let's look at a compare task together." Show students the task with the numbers covered up with sticky notes (see Figure 7.6).

Figure 7.6 Fishing Task With Numbers Covered

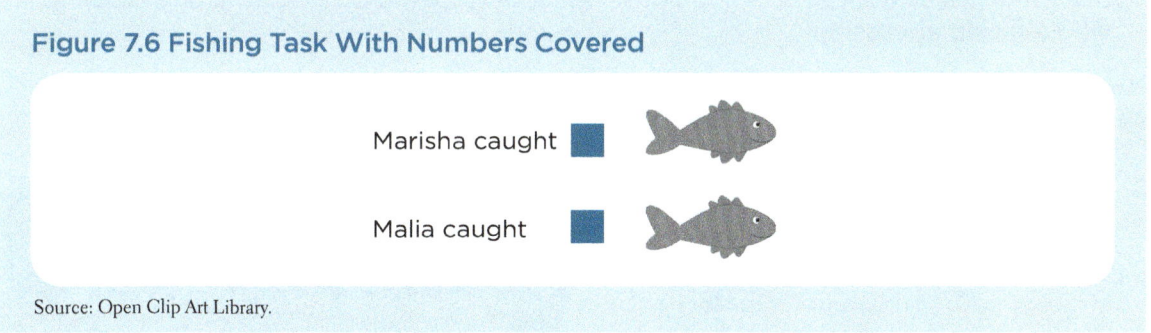

Source: Open Clip Art Library.

Each said they caught the most fish. Help them figure out who caught more.

8. Say, "What do you notice?" Elicit from the students that they need to know how many fish were caught to be able to compare.

9. Next, pull off the sticky notes and reveal the numbers 4 and 6. Ask, "How can you find out who caught more fish? Turn and talk with a partner to discuss your ideas."

10. **Observe** students as they talk with a partner. Listen with curiosity as they settle on and use a strategy.

11. Ask, "Who caught more fish? How do you know?" Select students to share the strategies they thought of with the class. Some strategies might include number lines and making sets using cubes.

12. Have the partners solve and observe. Some strategies students might suggest could include the following:

   » Students use a number line to compare the numbers. Students circle both numbers on the number line and determine that when looking at the numbers to the right of 0, the number that is the farthest distance away from 0 is the greatest or the number that is closest to 0 is the least.

   » Students might choose to count up from 1 and conclude that the last number stated in the sequence is the greatest number.

ACCESS AND EQUITY

If needed, provide counters for comparing numbers using a one-to-one strategy to support students who need additional support with comparing quantities. It may be necessary to suggest and/or have students model multiple strategies to help them transition from comparing the number of objects to comparing with numbers only.

Note: Consider using the Observation tool (see Appendix B).

FACILITATE

1. Read the task (Figure 7.3) to students:

   Marisha and Malia went fishing. Marisha caught 5 fish and Malia caught 8 fish. Each said they caught the most fish. Who caught more fish?

2. Show students number cards to represent the amount of fish. Say, "Tell how you know who caught the most fish. Work with your partner to solve this task." Make sure that students have access to tools such as number lines, number paths, and/or counters.

3. **Observe** students as they work and use the **Show Me** strategy to see and hear about the strategies students use. As you observe, note students' strategies to share during whole-group discussion.

Note: Consider using the Observation and Show Me tools for monitoring and recording student responses (see Appendix B).

4. When students have solved, say, "Now you and your partner will find a set of partners to make a *Pair-to-Pair Share.* Share your answer and your strategy with your group." Observe the student groups as they respond to the following:

   » What was the same about your strategies?

   » What was different about your strategies?

STRENGTHS SPOTTING

Incorporate questions like these to cultivate strengths building in students: "What is a challenge you faced while solving the problem? How do you think this strategy will help you tomorrow?" Recognize students when they exhibit these strengths.

Note: Consider using the Observation, Interview, and Show Me tools (see Appendix B).

CLOSE: MAKE THE MATH VISIBLE

1. Bring the students to a whole-group discussion. Ask, "Who caught more fish, Marisha or Malia? How do you know?" Students should state that Malia caught more fish. ("8 is more than 6.") Allow time for students to share their strategies. Refer to your monitoring notes to select students to share their work.

2. Ask the class the following **Hinge Question:** "If Nestor caught 4 fish and Bryce caught 9 fish, did Nestor catch more fish or fewer fish than Bryce? How do you know?"

3. **Extension Activity: Compare It!** For further practice with comparing numbers, provide students with two sets of number cards and allow students to play this comparing game that is similar to the traditional card game of War. Students play the game with a partner. Students place the number cards face down in a stack and take turns turning over the top card. Students compare the numbers written on the cards and decide which number is greater than the other. The student with the card that represents the greater quantity gets to keep both cards. At the end of the game, the student with the most cards is the winner. Students can also play the game where the student with the card that represents the smaller quantity gets to keep both cards. If students both draw the same number, they should place another set of cards on top—winner takes all. As student pairs play the game, make note of which students can easily compare the numbers without the use of additional tools. Provide appropriate scaffolds (number lines, number paths, counters, etc.) as needed.

TASK 13: THE FISHING CONTEST STUDENT PAGES

online resources — To download printable resources for this task, visit **resources.corwin.com/ ClassroomReadyMath/K-1**

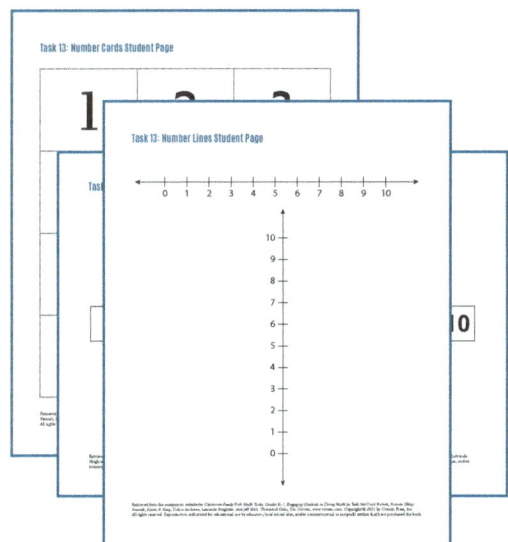

POST-TASK NOTES: REFLECTION & NEXT STEPS

CHAPTER

8

Operations and Algebraic Thinking

Addition and Subtraction

TASK 14: KINDERGARTEN: WHAT HAPPENS WHEN SOMEONE GIVES YOU CARROTS?

Represent addition and subtraction with objects, fingers, mental images, drawings, sounds (e.g., claps), acting out situations, verbal explanations, expressions, or equations.

Solve addition and subtraction word problems, and add and subtract within 10, e.g., by using objects or drawings to represent the problem.

TASK 15: GRADE 1: ARE THESE TWO PROBLEMS THE SAME?

Use addition and subtraction within 20 to solve word problems involving situations of adding to, taking from, putting together, taking apart, and comparing, with unknowns in all positions, e.g., by using objects, drawings, and equations with a symbol for the unknown number to represent the problem.

TASK 16: GRADE 1: COME TO THE PLAYGROUND

Solve word problems that call for addition of three whole numbers whose sum is less than or equal to 20, e.g., by using objects, drawings, and equations with a symbol for the unknown number to represent the problem.

Add and subtract within 20, demonstrating fluency for addition and subtraction within 10. Use strategies such as counting on; making ten (e.g., $8 + 6 = 8 + 2 + 4 = 10 + 4 = 14$); decomposing a number leading to a ten (e.g., $13 - 4 = 13 - 3 - 1 = 10 - 1 = 9$); using the relationship between addition and subtraction (e.g., knowing that $8 + 4 = 12$, one knows $12 - 8 = 4$); and creating equivalent but easier or known sums (e.g., adding $6 + 7$ by creating the known equivalent $6 + 6 + 1 = 12 + 1 = 13$).

TASK 17: GRADE 1: FROGS JUMPING AWAY

Add and subtract within 20, demonstrating fluency for addition and subtraction within 10. Use strategies such as counting on; making ten (e.g., $8 + 6 = 8 + 2 + 4 = 10 + 4 = 14$); decomposing a number leading to a ten (e.g., $13 - 4 = 13 - 3 - 1 = 10 - 1 = 9$); using the relationship between addition and subtraction and creating equivalent but easier or known sums (e.g., adding $6 + 7$ by creating the known equivalent $6 + 6 + 1 = 12 + 1 = 13$).

TASK 18: GRADE 1: COMING AND GOING

Add and subtract within 20, demonstrating fluency for addition and subtraction within 10. Use strategies such as counting on; making ten (e.g., $8 + 6 = 8 + 2 + 4 = 10 + 4 = 14$); decomposing a number leading to a ten (e.g., $13 - 4 = 13 - 3 - 1 = 10 - 1 = 9$); using the relationship between addition and subtraction and creating equivalent but easier or known sums (e.g., adding $6 + 7$ by creating the known equivalent $6 + 6 + 1 = 12 + 1 = 13$).

TASK 19: GRADE 1: GOING HOME AFTER SCHOOL

Use addition and subtraction within 20 to solve word problems involving situations of adding to, taking from, putting together, taking apart, and comparing, with unknowns in all positions, e.g., by using objects, drawings, and equations with a symbol for the unknown number to represent the problem.

Determine the unknown whole number in an addition or subtraction equation relating three whole numbers. For example, determine the unknown number that makes the equation true in each of the equations 8 + ? = 11, 5 = ___ – 3, 6 + 6 = ___.

Anticipating Student Thinking: Early mathematics learning opportunities involving counting naturally extend to comparisons (see the previous chapter) and then to combining, joining, separating, and comparing amounts. Task-based foundational activities involving addition and subtraction start right here! The first task is at the kindergarten level (Task 14), all others address standards that emphasize adding and subtracting within 20. The chapter's tasks will engage your students in the development of important foundational components of addition and subtraction. These include solving word problems that involve addition and subtraction situations (e.g., adding to, taking from, putting together, taking apart, and comparing with unknowns), and using strategies such as counting on, making a 10, decomposing numbers, and the relationship between addition and subtraction. All of the tasks will involve students in using representations as they add or subtract.

THINK ABOUT IT

This chapter's focus involving beginning experiences with addition and subtraction, including the addition and subtraction situations, will engage your students in doing math tasks where they may struggle a bit. View such productive struggle as an opportunity for them to take the time to make sense of the mathematics they are learning. Support your students as they persevere, and make sure to provide opportunities to share their solution journeys.

Task 14

What Happens When Someone Gives You Carrots?

Solve addition word problems

TASK

What Happens When Someone Gives You Carrots?

Figure 8.1 Carrots

Source: Amin Yusifov/iStock.com

Beth has four carrots (see Figure 8.1). If her friend gives her more carrots, does Beth have more or fewer carrots than she did at the start?

TASK PREPARATION

- Ensure that students have access to counters.

- Organize the students into pairs toward the end of the Launch, when discussing the problem presented to the whole group regarding Beth's four carrots.

- Provide carrot cards to the student pairs for the Facilitate portion of the task lesson.

LAUNCH

1. Bring the class together and introduce the task by asking the students, "Do you like to eat carrots?"

2. Show four of the carrot cards and ask, "Which one doesn't belong? Turn and talk with a partner and decide which one doesn't belong. Be prepared to explain which one doesn't belong and tell your reasons why." Allow time for student pairs to work together. Students will likely focus on how many carrots there are on each card. The main idea here is to introduce the students to the carrot cards and to give them initial exposure to Which One Doesn't Belong? questions.

3. Bring the class together for whole-group discussion and encourage students to share their thinking.

Mathematics Standards

- Represent addition and subtraction with objects, fingers, mental images, drawings, sounds (e.g., claps), acting out situations, verbal explanations, expressions, or equations.

- Solve addition and subtraction word problems, and add and subtract within 10, e.g., by using objects or drawings to represent the problem.

Mathematical Practices

- Make sense of problems and persevere in solving them.

- Model with mathematics.

Vocabulary

- more
- fewer
- less, less than
- numbers one to ten
- add to

Materials

- counters
- Carrot Cards student page (cards include numbers from 1 to 6), one set for each pair
- whiteboard
- marker

4. Say, "Class, Beth has four carrots." Show the carrot card that has four carrots on it.

5. Ask, "Could someone please come up and count the carrots on the card?" Have a student come up and count the four carrots.

6. Say, "In today's task, we are going to use these carrot cards. Each card has a different number of carrots on it. Here's the task that I would like you all to start with: Beth has four carrots. If her friend gives her more carrots, does Beth have more or fewer carrots than she did at the start? Work with your partner and discuss if you think Beth now has more carrots or fewer carrots than she did at the start."

7. Record on the whiteboard the number of student pairs that think Beth has more carrots and the number of student pairs that think Beth has fewer carrots.

ACCESS AND EQUITY

Make sure that you vary the partners so that all students have a voice in the classroom. Monitor conversations to notice which students might need more opportunities to share their ideas.

STRENGTHS SPOTTING

Match partners who demonstrate different strengths and let students know that they each have value and strengths to contribute to the partnership.

FACILITATE

1. Provide carrot cards and counters to each of the student pairs. Then say, "Now, to help us figure out if Beth has more or fewer carrots, each team should pick one carrot card from your pile. This card is the number of carrots that Beth's friend gives her. With your partner, use the counters to show the 4 carrots Beth had to start with, and then show counters for the number of carrots Beth's friend gives her. Now we can figure out how many carrots Beth has now. Is it more than or fewer than the 4 carrots that Beth started with?"

2. **Observe** the student pairs as they represent the number of carrots using the counters. Ask them to discuss their strategy for deciding how many they are adding to 4, and the new amount.

3. As you observe the student pairs, consider asking the following interview questions:

   » How did you figure out how many carrots Beth had at the end?

   » Did you count? What numbers did you use?

   » Could you write a number sentence for this problem?

   » Was your final answer more or fewer than 4? Was that the same or different from how you voted on the whiteboard?

STRENGTHS SPOTTING

Students will need time and opportunities to work with their peers and freedom to explore their thinking through opportunities to share initial ideas, sometimes referred to as rough-draft talk.

Note: Consider using the Observation and Interview tools for recording pair responses (see Appendix B).

CLOSE: MAKE THE MATH VISIBLE

1. Bring the class together and have the pairs share their solution strategies and number sentences. Note that the number sentences will be slightly different because of the various carrot cards chosen. Focus on how the teams solved the problems and on their conclusion that Beth ended up with more carrots.

2. For each pair that shares, write on the board the number of carrots that were used in the problem (e.g., Beth's 4 and 5 more); a description or picture of the solution strategy the students used ("We put both cards in front of us and we counted the carrots 1, 2, 3, . . . 9"); and the conclusion (Beth has more carrots).

3. After several pairs have shared, the board should have many different "add to" values, with multiple strategies, and all with the conclusion that Beth ended up with more carrots.

4. Bring the class discussion back to the initial predictions/votes on the whiteboard. Ask the class if their initial predictions were correct or not, given the examples/evidence on the board.

TASK 14: CARROT CARDS

 To download printable resources for this task, visit **resources.corwin.com/ ClassroomReadyMath/K-1**

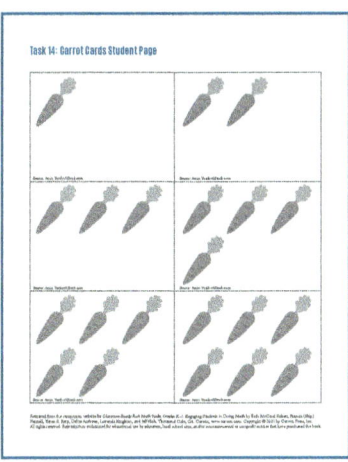

POST-TASK NOTES: REFLECTION & NEXT STEPS

Mathematics Standard

- Use addition and subtraction within 20 to solve word problems involving situations of adding to, taking from, putting together, taking apart, and comparing, with unknowns in all positions, e.g., by using objects, drawings, and equations with a symbol for the unknown number to represent the problem.

Mathematical Practices

- Make sense of problems and persevere in solving them.
- Reason abstractly and quantitatively.

Vocabulary

- add
- put together
- join
- change

Materials

- counters—at least 10 in a baggie if possible
- Rock Cards student pages (cut into sets), one set for each pair

Task 15

Are These Two Problems the Same?

Solve Add to, Change Unknown word problems

TASK

Are These Two Problems the Same?

Figure 8.2 A Rock Collection

Source: Pixabay.com and Pobytov/iStock.com

- Tristan has 5 rocks (see Figure 8.2). His mother gives him some more rocks. Now Tristan has 9 rocks. How many rocks did Tristan's mother give him?

- Maria has 5 rocks in her collection. How many more rocks does Maria need to find so that she has 9 rocks in her collection?

TASK PREPARATION

- In this task, students will work with two Add to, Change Unknown problems that are worded slightly differently. The aim of the task is for students to see that while the words are different in each problem, the underlying mathematical structure is the same. They will also have the opportunity to solve the two problems.

ACCESS AND EQUITY

The rock cards are optional and allow for student flexibility when solving the two problems in the task. Some students can benefit from the use of a pictorial version of the 5-group as they use counting-on strategies to solve the problems.

- Have students work in pairs. Ensure that there are either containers that contain sufficient manipulatives for student pairs to share, or prepare baggies of counters that include at least 10 counters in each bag.

ALTERNATE LEARNING ENVIRONMENT

Students at home can be encouraged to collect items around the house (e.g., rocks from the yard, toy blocks, etc.) to use as manipulatives.

LAUNCH

1. Project the following numberless word problem so that the whole class can see it: "Maria has some rocks in her collection. She wants to have more rocks in her collection." Ask students what they think about when they hear those sentences: "Do any of you have collections of your own? What do you collect? How many do you have? How many rocks do you think Maria has?"

2. Post the following problem next to the numberless problem from Launch step 1:

   » Maria has 5 rocks in her collection. How many more rocks does Maria need to find so that she has 9 rocks in her collection?

3. Ask the students to *Turn and Talk* with their neighbor about these questions: "What has changed in the problem? What new information do we have? What did we learn from this new information?"

4. Reveal the second problem and compare it to the Maria problem:

   » Tristan has 5 rocks. His mother gives him some more rocks. Now Tristan has 9 rocks. How many rocks did Tristan's mother give him?

5. Have students work with their partner again to answer the following questions: "How are these two problems the same? How are they different?"

6. Record students' ideas.

FACILITATE

1. Make sure students have access to counters.

2. **Observe.** Walk around the class and ask students/teams to clarify the strategies they are discussing. Pay attention to which problem they are working on. Are they able to solve both problems?

3. **Interview.** As you observe the progress of the student pairs, select individual students or pairs to interview. Consider asking:

   » How did you solve each problem?

   » Did you count on? What numbers did you use?

   » How are the two problems the same?

   » How are they different?

! PRODUCTIVE STRUGGLE

Students who need experience with Add to, Change Unknown problems may only have had prior experience with Result Unknown problems. They should be encouraged to model the action in the problem. What happens in the story? Can we count out Maria's rocks? Then what happens? Maria's rock collection problem will likely be easier for them to solve, so direct them to work with it if they began with Tristan's problem.

» What does the 5 represent in each problem?

» How would you write an equation for each of the problems?

Note: Consider using the Observation and Interview tools (individual or small group) to provide a record of what you observe and interview comments (see Appendix B).

CLOSE: MAKE THE MATH VISIBLE

1. Bring the class back together for a whole-class discussion about the two word problems, asking the class to discuss how the two problems are similar and different. The students may focus on the problem contexts—that both problems are about rocks. They could also share that in each case, Tristan and Maria start and end up with the same number of rocks. They might notice that Tristan's mother is in the first problem, while Maria finds rocks on her own.

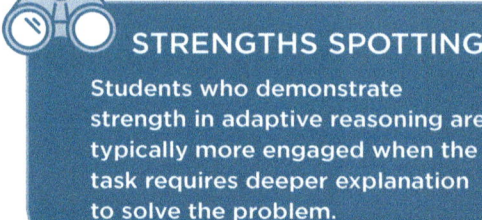

STRENGTHS SPOTTING

Students who demonstrate strength in adaptive reasoning are typically more engaged when the task requires deeper explanation to solve the problem.

2. From your earlier interview conversations with the student pairs, discuss different topics that could differentiate the two problems. For example, you could draw on discussions of the structure of the two problems being slightly different or that the addend in the two problems is slightly different because of Tristan's mother giving some rocks instead of Maria finding more rocks. Ask, "Was one of these problems more challenging for you to solve, or were they the same? Why do you feel like one problem is harder than the other?" Select pairs that believed that one problem was harder than the other to share their thinking. Ask, "Can you write an equation to describe what's happening in each problem?"

3. Close the task by having the students compare the equations of the two problems. Reiterate that while word problems can be worded differently and the stories could be similar, the equations can be the same.

TASK 15: ROCK CARDS STUDENT PAGES

To download printable resources for this task, visit **resources.corwin.com/ ClassroomReadyMath/K-1**

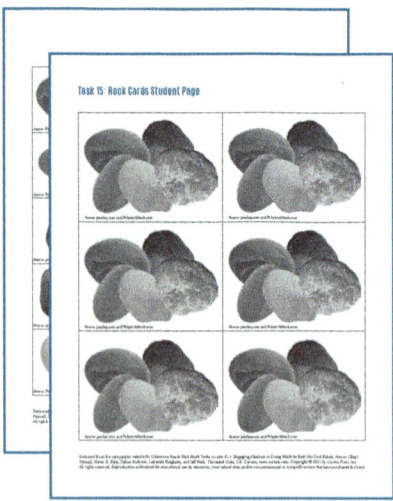

POST-TASK NOTES: REFLECTION & NEXT STEPS

Task 16
Come to the Playground

Solve addition word problems with three addends

TASK

Come to the Playground!

Figure 8.3 The Playground

Source: robuart/iStock.com

There were 4 children on the playground. Soon, 5 more children came to the playground. Then 6 more children arrived at the playground. How many children are on the playground (Figure 8.3) now?

TASK PREPARATION

- This task involves the Add to, Result Unknown addition situation, with examples that facilitate students' use of making tens.

- Provide students with counting materials.

- Because there are three addends, you should be mindful about using the numbers that your students are comfortable with.

- Students should solve the task independently, and then compare equations and share solution strategies in pairs or in small groups.

- When writing problems with the Add to, Change Unknown situations, make sure that the word problem includes a specific action. Explicit actions help the student model the problem.

- Building off of Add to, Result Unknown situations, the inclusion of another addend allows students to explore the associative property and add the number of children on the playground in multiple ways.

- When the students experience problems with three addends, anticipate their solution strategies. Will they add the three numbers in the order that they are given in the problem (e.g., $4 + 5 + 6$), or, if they are becoming more flexible with their thinking, will they make a 10 with the 4 and the 6 and then add 5 to get 15.

LAUNCH

1. Bring the class together as a whole group and have a discussion about situations that happen during recess and after lunch, when some children go to the playground first and then more children join them.

2. Then, either go to the playground or designate a spot in the classroom as the playground.

3. Say, "The math problem that we are going to start with today is this: There were 4 children on the playground. Soon, 5 more children came to the playground. Then 6 more children arrived at the playground. How many children are on the playground now?"

4. Say, "Let's act it out together! How many children start on the playground? (4) So I need four volunteers to come on to the playground. Then what happens in the story? Five more children come to the playground. So what should we do now? (Count out the five children that move over to the playground.) Then what happens? Six more children arrive at the playground! (Count out the six additional children that go to the playground). How can we find out how many children are on the playground now?" (Count out all of the children on the playground from 1 to 15.)

5. Return to the classroom and present the following problem on the board:

   > There were _____ children on the playground. Soon, _____ more children came to the playground. Then _____ more children arrived at the playground. How many children are on the playground now?

6. Read the problem to the class and say, "We solved a problem that had 4, 5, and 6 children at the playground. For this problem, pick three cards from your number cards. Those are the numbers of children that arrive at your playground."

ACCESS AND EQUITY

The focus of this task is to solve the problem, not to read it. Eliminate a potential barrier to task access by reading the problem and repeating it if needed.

FACILITATE

1. Make sure students have access to all of their mathematics tools, such as counters, ten frames, fingers, and things to draw with. Have the students solve the task in multiple ways if possible, sharing their strategies with at least one other student.

Vocabulary

- add
- add to
- combine
- join
- sum
- model
- equation
- numbers 1 to 20

Materials

- counters
- ten frames (optional)
- Number Cards student page (1–10, cut into sets), one set per student
- Come to the Playground student page, one per student

STRENGTHS SPOTTING

Sharing solutions publicly provides students with opportunities to share developmentally appropriate strategies with one another to promote students' solution pathways.

2. **Observe** the students as they work. Use the **Show Me** technique to see and hear about the strategies used by the students. Focus on what numbers the students add first. Do they add the numbers in the order of the problem (adding the 4 and the 5 first and then adding on the 6), or do they make a 10 with the 4 and 6 and then add on the 5? They might also add the numbers in reverse order.

3. Using selected **Show Me** responses, display equations used for the task's solution.

Note: Consider using the Observation and Show Me tools for recording student responses and considering student strategies to be presented in the Close portion of the task lesson (see Appendix B).

CLOSE: MAKE THE MATH VISIBLE

1. Bring the class together and have students share their equations and the strategies they used to solve the playground problem.

2. Highlight one of the student strategies where the student makes a 10 with the 4 and the 6, and then adds 5.

3. **Hinge Question.** There were 5 students who went to the playground after school, then 3 more came, and then 5 more. Were there more than 10 children at the playground? How do you know?

4. Student responses to the "How do you know?" part of the Hinge Question should reveal student confidence and understanding regarding the use of the make ten strategy.

TASK 16: COME TO THE PLAYGROUND STUDENT PAGES

 To download printable resources for this task, visit **resources.corwin.com/ ClassroomReadyMath/K-1**

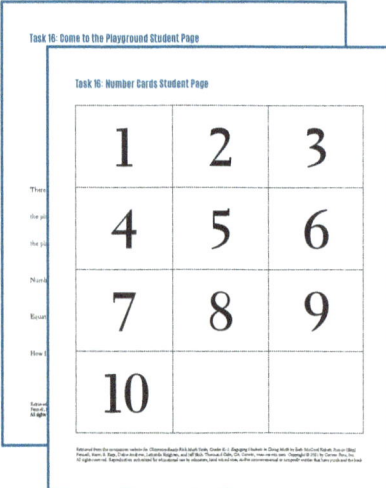

Mathematics Standard

- Add and subtract within 20, demonstrating fluency for addition and subtraction within 10. Use strategies such as counting on; making ten (e.g., 8 + 6 = 8 + 2 + 4 = 10 + 4 = 14); decomposing a number leading to a ten (e.g., 13 − 4 = 13 − 3 − 1 = 10 − 1 = 9); using the relationship between addition and subtraction (e.g., knowing that 8 + 4 = 12, one knows 12 − 8 = 4); and creating equivalent but easier or known sums (e.g., adding 6 + 7 by creating the known equivalent 6 + 6 + 1 = 12 + 1 = 13).

Mathematical Practice

- Make sense of problems and persevere in solving them

Vocabulary

- mathematical operation
- subtract
- take away

Materials

- counters
- little ten frames (optional, one for each student)
- Number Cards student page (1–10)
- Frogs on a Log student page
- Frogs on a Log Frog Counters (cut into sets), one set for each student

Task 17
Frogs Jumping Away

Solve Take From word problems as numberless word problems

TASK

Frogs Jumping Away

There were 6 frogs sitting on the log (see Figure 8.4). Then 2 frogs jumped away. How many frogs are sitting on the log now?

Figure 8.4 Frogs on a Log

Source: Log by Oqvector/iStock.com; Frogs by Angyee054/iStock.com

TASK PREPARATION

- Numberless word problems provide opportunities for students to better understand the structure of word problems. They can be used for many different word problem types; this task uses the Take From, Result Unknown subtraction situation.

 **STRENGTHS SPOTTING**

Highlight the sense making and reasoning that the students use as they make sense of the numberless word problems. Help students develop their points of power so that they see that they can solve word problems and have a positive attitude toward mathematics.

- Provide students with counters.

LAUNCH

1. Gather the class together and announce the following: "Today in math, we are going to talk about what we notice. Then we will move on to the next picture and see what has changed. We will keep doing this until we end up with word problem, which you will all get to solve!"

2. Say, "There were some frogs sitting on the log. Some frogs jumped away. How many frogs are left?" (To class) "What happened here? Can you imagine this situation in your mind? Tell me about this picture. What might this picture look like?" Focus the class on the action in the

story. You want the students to understand what is happening in the story: that there are frogs jumping away from the log.

3. Show the picture and say: "There were 6 frogs sitting on the log. Some frogs jumped away." The difference between this problem and the initial problem is that there are 6 frogs specified at the beginning of the problem.

4. Ask, "What changed between the first story and this story? How many frogs do you think jumped away?" If appropriate, ask, "What would an equation look like for this story?"

FACILITATE

1. Say, "There were 6 frogs sitting on the log. Then 2 frogs jumped away." How many frogs are sitting on the log now?

2. Make sure students have access to their mathematical tools (counters, ten frame, drawing materials). Have them solve the problem in multiple ways, sharing their strategies with at least one other student.

3. **Observe** the students as they work. Ask, "Show me how you solved the problem."

4. When the students have solved the first problem, have them draw number cards to change the numbers in the problem.

> ### ❗ PRODUCTIVE STRUGGLE
>
> As with other word problem types, students may not be able to generate multiple strategies. Celebrate the successful use of counters. Pair students who are using counting-back strategies to highlight multiple strategies.

Note: Consider using the Observation and Show Me tools for recording and organizing student responses. This may be particularly helpful for the student sharing in the Close portion of the task lesson (see Appendix B).

CLOSE: MAKE THE MATH VISIBLE

1. Bring the class together and have students share specific strategies they used for the 6 frogs sitting and 2 frogs jumping away task.

2. Try to arrange for students who used the same strategy but with different representations to share how their strategies were both alike and different. For example, some students might directly model with counters. Others might have drawn a picture. Encourage such multiple representations.

3. Also, call on students who used different solution strategies to share their work.

TASK 17: FROGS JUMPING AWAY STUDENT PAGES

To download printable resources for this task, visit **resources.corwin.com/ ClassroomReadyMath/K-1**

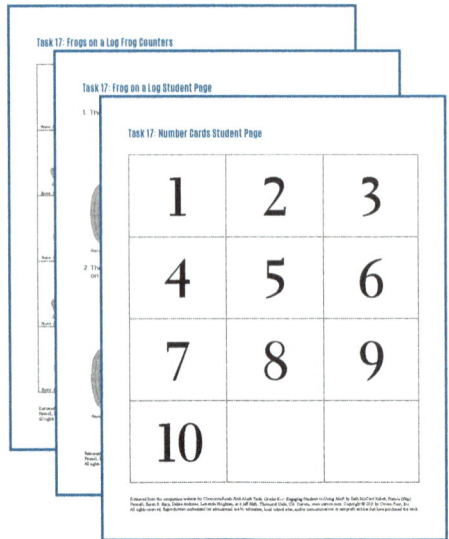

POST-TASK NOTES: REFLECTION & NEXT STEPS

Task 18
Coming and Going

Explore the inverse relationship between addition and subtraction

TASK

Coming and Going

Figure 8.5 The Park

Source: pixabay.com

a. There were 11 children at the park. Then 5 of the children went home. How many children were still at the park?

b. There were 6 children at the park. Then 5 more children came to the park. How many children are at the park now?

TASK PREPARATION

- This task pairs two word problems with similar contexts (see Figure 8.5) to provide an opportunity for students to see the inverse relationship between subtraction and addition.

- Ensure that there are either containers that contain sufficient manipulatives for students to share, or prepare baggies of counters that include at least 11 counters in each bag.

ALTERNATE LEARNING ENVIRONMENT

Students could use virtual counting cubes to provide access to manipulatives in a remote learning setting: www.didax.com/apps/unifix/

- Assemble the whole class for the Launch. Then have them work in small groups for the Facilitate and Close portions of the task lesson.

Mathematics Standard

- Add and subtract within 20, demonstrating fluency for addition and subtraction within 10. Use strategies such as counting on; making ten (e.g., $8 + 6 = 8 + 2 + 4 = 10 + 4 = 14$); decomposing a number leading to a ten (e.g., $13 - 4 = 13 - 3 - 1 = 10 - 1 = 9$); using the relationship between addition and subtraction (e.g., knowing that $8 + 4 = 12$, one knows $12 - 8 = 4$); and creating equivalent but easier or known sums (e.g., adding $6 + 7$ by creating the known equivalent $6 + 6 + 1 = 12 + 1 = 13$).

Mathematical Practices

- Make sense of problems and persevere in solving them.
- Look for and make use of structure.

Vocabulary

- subtract
- take away
- add; add to
- equal sign
- equation

Materials

- counters in baggies
- whiteboards
- Playground student page, one per student
- Stick Figures student page, one per student

LAUNCH

1. Say, "Today, we are going to talk about coming and going. Can anyone think of examples of when you come and when you go?"

2. Choose and discuss one of the student contexts (e.g., coming from and going to school). Then, have students act out a coming and going action in front of the class. For example, "Five brilliant mathematicians came to the front of the class this morning." Have five students come up to the front of the room. "Three brilliant mathematicians then went back to their seats." Have three of the five students return to their seats.

3. Ask, "What would be an example of an equation to represent what just happened?" Discuss the five students as the initial number of mathematicians in the front of the room and the mathematical operation for the students returning to their seats.

4. Discuss the situation, emphasizing what happened when the mathematicians "came" and "went."

5. Repeat the acting-out process with five initial students and then four more students joining them. Say, "Five brilliant mathematicians came to the front of the class this morning." Have five students come up to the front of the room. "Four more brilliant mathematicians then joined them in the front of the class."

FACILITATE

1. Working in small groups, have students act out both task problems and write an equation on shared group whiteboards to represent both "going" and "coming":

   a. There were 11 children at the park. Then 5 of the kids went home. How many kids were still at the park?

   b. There were 6 kids at the park. Then 5 more kids came to the park. How many kids are at the park now?

   Say, "Work with your group members and write down an equation to represent each problem. For these two problems, which one has the most children at the park in the end? How do you know?"

> **! PRODUCTIVE STRUGGLE**
>
> In order for students to have opportunities to make sense of tasks, teachers must resist the urge to intercede too soon. Allow time for students to make sense of the problem. Some students might be comfortable writing the numbers from the problems but have not yet had experience working with the addition, subtraction, and equal signs. Normalize the struggle by letting students know that all mathematicians struggle. When students experience success, focus on the good feelings that productive struggle produces!

3. As students work, monitor their progress using Interview and Show Me techniques.

   • **Interview:**

     » How did you solve the problem?

     » Tell me why you solved it that way. What else can you tell me about what you did?

     » What do the numbers represent in each of the equations? How did you decide which operation to use?

     » With these two problems, what is the same? What is different?

- **Show Me.** Show me your equations for your "going" problem. Show me the mathematical symbol for your "coming" problem.

Note: Consider using the Interview (individual or small group) and Show Me tools (see Appendix B) for recording comments and noting student performance as needed.

4. After completing one pair of problems, encourage the student teams to try "going" and "coming" problems again but with different numbers (e.g., 13, 7, and 6).

CLOSE: MAKE THE MATH VISIBLE

1. Facilitate a **Notice and Wonder** Gallery Walk so that students can see everyone's equations for their own word problems represented.

2. **Hinge Question.** Ask students what they notice about all of the "going" and "coming" equations.

3. Then, highlight what happens mathematically in all of the "going" (subtraction) problems as well as the "coming" (addition) problems.

STRENGTHS SPOTTING

Provide students an opportunity to recognize the contributions that their classmates offered. When students provide strengths-based feedback to their peers, students are more likely to take risks and innovate. Ask students questions such as "What do you notice about the equations?"

4. Point out how every time the number of children who leave matches the number of children who arrive, you end up with related equations.

5. Ask teams to share their thoughts about the relationship between the two equations.

6. Have them try to generalize the relationship between 11 − 5 = 6 and 6 + 5 = 11, again emphasizing what each of the numbers represents in both of the equations.

TASK 18: COMING AND GOING STUDENT PAGES

 To download printable resources for this task, visit **resources.corwin.com/ ClassroomReadyMath/K-1**

Task 19
Going Home After School

Solve Take From, Change Unknown problems

TASK

Going Home After School

Figure 8.6 Students Leaving School

Source: School by mutsMaks/iStock.com; Children by Sudowoodo/iStock.com

There were 12 students at school at the end of the day. Some of them went home (see Figure 8.6). Now there are 3 students left at school. How many students went home?

TASK PREPARATION

- This task focuses on the introduction of a specific addition/subtraction situation called Take From, Change Unknown.

- This problem situation is significant as it is typically the most difficult of the four Add to, Take From problem situations introduced at this grade level. However, research on student thinking demonstrates that children who have had their own thinking valued when solving the other three problem types are able to successfully solve problems involving this situation without direct instruction.

- This task engages the class in a context-connected problem that culminates in students generating the appropriate equation to fit the problem's addition/subtraction situation (Add to and Take From, Change Unknown). It is assumed that students have had prior experience with Add to and Take From/Result Unknown and Start Unknown problem types (2 + 4 = ?, 6 − 2 = ?; ? + 2 = 6, ? − 2 = 4).

Mathematics Standards

- Use addition and subtraction within 20 to solve word problems involving situations of adding to, taking from, putting together, taking apart, and comparing, with unknowns in all positions, e.g., by using objects, drawings, and equations with a symbol for the unknown number to represent the problem.

- Determine the unknown whole number in an addition or subtraction equation relating three whole numbers. For example, determine the unknown number that makes the equation true in each of the equations $8 + ? = 11$, $5 = \underline{\quad} − 3$, $6 + 6 = \underline{\quad}$.

Mathematical Practices

- Make sense of problems and persevere in solving them.
- Model with mathematics.

Vocabulary

- subtract
- take from
- model
- equation
- equal sign
- numbers 1 to 12

Materials

- counters
- Going Home After School Problem Cards for Centers student pages

- It is also expected that the students have had experiences with counting up, counting backward, and writing equations for the addition/subtraction situations noted above.

- To focus on sense making, the subtraction situation (Change Unknown) presented by the teacher needs to mirror the task.

- The whole-group implementation of the task within the Launch will require space for a group of 12 students (in a group) to move around.

- Students will need access to the counting materials noted earlier.

LAUNCH

1. Bring the class together as a whole group and have a brief discussion about the following question: What happens when the school day ends? (students leave school).

2. Discuss how we can use the end of the school day as part of a math problem.

3. Now, present the task: There were 12 students at school at the end of the day. Some of them went home. Now there are 3 students left at school. How many students went home?

4. As this is the introduction to this problem situation, the task will be modeled using actual students. Ask the class, "How does this problem start? What should we do to start solving it? Can you imagine this situation?"

5. When someone suggests to count out 12 counters, respond by saying, "Today we will use classmates as counters!" Choose 12 students to stand up and move to one side of the room: "These are the 12 students at school at the end of the day. Now what happens in the story?"

6. When you get the response that students need to go home, ask the class, "What should we have these 12 students do?" Elicit the response that the students need to go home one at a time.

7. "How do you know when to stop?" Get the student response that the problem says "until there are 3 left."

8. From the group of 12 students, have them "go home" one at a time by walking to the other side of the classroom. After several of them have "gone home," ask the class if more students need to keep going and how they know.

9. When there are three students left "at school," ask the class, "How many students went home? How do you know?"

10. With the entire class, work on the development of the equation that represents the task's problem.

   » Ask, "How many students were at school at the end of the day?" Elicit responses from the students. For the questions in this section, everything should be student driven and not come from the teacher. Say, "So, we can write down a '12.'"

   » Say, "Then, they went home. What mathematical symbol can we use for when we are taking away from this 12? A subtraction sign!"

   » Ask, "When we first started solving the problem, did we know how many students went home? No. That was our unknown. We will show where the unknown number will go with a question mark."

» Continue by asking, "And how many students were left at school? 3. So, we write 12 – ? = 3."

» Ask, "And when we solved the problem, what number does the question mark represent? How many students went home? 9! So, 12 – 9 = 3."

FACILITATE

1. Create four centers, each of which contains a different problem:

   » There were 9 students at school at the end of the day. Some of them went home. Now there are 2 students left at school. How many students went home?

   » Antonio had 11 rocks in his rock collection. He gave some rocks to his sister. Now he has 6 rocks left. How many rocks did he give to his sister?

   » There were 8 children playing soccer on the field. Some left to go inside. Now there are 3 children still on the field. How many children went inside?

   » TJ had 7 cookies. At snack, she ate some of them. Now she has 4 cookies left. How many cookies did she eat?

2. Have the students solve the problem in each center in multiple ways, sharing their strategies with at least one other student and recording their solution on a poster found in each center. Have students move on to a different center after completing each problem.

3. **Observe** students as they work, using the Interview (small group) and Show Me techniques to engage students in discussing their solution strategies.

4. Encourage the students to write the equation for the problem they just solved.

Note: Consider using the Observation, Interview (small group), and Show Me tools to make note of the level of student involvement and record student solution strategies for the task (see Appendix B).

> ### ! PRODUCTIVE STRUGGLE
>
> As some students may not be familiar with this word problem type, have them focus on the specific action in the problem, similar to the Launch. Students should not feel pressure to get to every center but rather spend time making sense of the problems.

> ### STRENGTHS SPOTTING
>
> Celebrate students' positive dispositional strengths as they listen to others' ideas and work cooperatively and collaboratively with other students. Consider highlighting the strengths that you see during the Close.

CLOSE: MAKE THE MATH VISIBLE

1. Bring the class together and have students share their strategies and the equation they used to represent the problem.

2. Highlight particularly interesting solution strategies. A commonly used strategy is when the student represents all of the people in the problem with a counter for each person. So for the first center problem, the student counts out 9 counting cubes, and then pulls them away one at a time until there are 2 left. The student then counts how many cubes were pulled away. Unless formally taught to do so, students do not usually count out 9 counting cubes and then pull away 2, as that is not what happens in the problem. Another solution strategy is when the student does not need to represent the first set of 9 counting cubes, but instead puts the number 9 in their head and counts backward,

raising one finger with each count: 8, 7, 6, 5, 4, 3, 2. The student then counts how many fingers are raised to get the answer: 7. A third strategy is when the student uses knowledge of a fact that they know to help them solve the problem: "I don't know 9 minus something equals 2, but I do know that 9 minus 6 equals 3. Take away 1 more and the answer is 7.

3. Announce to the class that they have now solved a new type of problem (Change Unknown)!

STRENGTHS SPOTTING

As students share their ideas, take the opportunity to notice how students listen as well as how and when they share. Focus on the ways that children show how they listen with their bodies and facial expressions. Curiosity and interest in others' ideas are important mathematical strengths to celebrate.

TASK 19: GOING HOME AFTER SCHOOL STUDENT PAGES

 To download printable resources for this task, visit **resources.corwin.com/ ClassroomReadyMath/K-1**

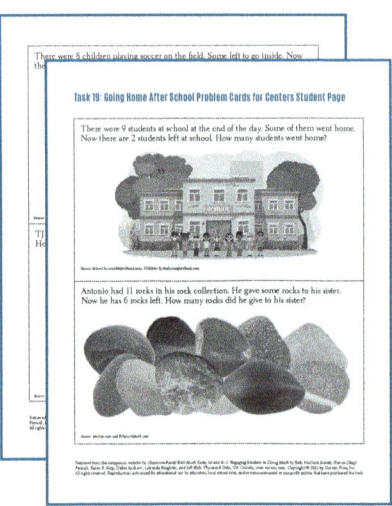

POST-TASK NOTES: REFLECTION & NEXT STEPS

Operations and Algebraic Thinking

Addition and Subtraction Relationships

TASK 20: GRADE 1: MAKE IT TRUE, MAKE IT FALSE

Understand the meaning of the equal sign, and determine if equations involving addition and subtraction are true or false. For example, which of the following equations are true and which are false? $6 = 6$, $7 = 8 - 1$, $5 + 2 = 2 + 5$, $4 + 1 = 5 + 2$.

TASK 21: KINDERGARTEN: SPLAT AND SPLIT

Decompose numbers less than or equal to 10 into pairs in more than one way, e.g., by using objects or drawings, and record each decomposition by a drawing or equation (e.g., $5 = 2 + 3$ and $5 = 4 + 1$).

TASK 22: GRADE 1: WHAT'S MY CARD?

Understand subtraction as an unknown-addend problem. *For example, subtract 10 – 8 by finding the number that makes 10 when added to 8.*

TASK 23: GRADE 1: LOTS OF EQUATIONS

Determine the unknown whole number in an addition or subtraction equation relating three whole numbers. For example, determine the unknown number that makes the equation true in each of the equations $8 + ? = 11$, $5 = \_\_ - 3$, $6 + 6 = \_\_$.

TASK 24: KINDERGARTEN: AT THE BOWLING ALLEY

For any number from 1 to 9, find the number that makes 10 when added to the given number, e.g., by using objects or drawings, and record the answer with a drawing or equation.

TASK 25: GRADE 1: KEEP MAKING TENS!

Add and subtract within 20, demonstrating fluency for addition and subtraction within 10. Use strategies such as counting on; making ten (e.g., $8 + 6 = 8 + 2 + 4 = 10 + 4 = 14$); decomposing a number leading to a ten (e.g., $13 - 4 = 13 - 3 - 1 = 10 - 1 = 9$); using the relationship between addition and subtraction (e.g., knowing that $8 + 4 = 12$, one knows $12 - 8 = 4$); and creating equivalent but easier or known sums (e.g., adding $6 + 7$ by creating the known equivalent $6 + 6 + 1 = 12 + 1 = 13$).

Anticipating Student Thinking:
This chapter's tasks will engage your students in literally applying addition and subtraction relationships as they begin to both develop and deepen conceptual understandings related to addition and subtraction and, importantly, a sense of number. This is so important. The kindergarten tasks (Tasks 21 and 24) emphasize decomposing numbers and making tens, respectively. The chapter's first-grade tasks emphasize strategies related to addition and subtraction and early work involving equations. Your students will be working in pairs or small groups and with a variety of representations as they engage the chapter's tasks. Classroom-based formative assessment techniques will be an important component of each task lesson's Facilitate or Close as they provide opportunities for you to monitor student and class progress and provide the feedback you will need to consider your next steps instructionally.

THINK ABOUT IT
As you close a task, think about how the responses provided by your students make their mathematics learning visible to you and them. Importantly, when the task lesson is completed, what's your takeaway? Find the time to use the task lesson template to record your reflection and your next steps.

Mathematics Standard

- Understand the meaning of the equal sign, and determine if equations involving addition and subtraction are true or false. For example, which of the following equations are true and which are false? $6 = 6$, $7 = 8 - 1$, $5 + 2 = 2 + 5$, $4 + 1 = 5 + 2$.

Mathematical Practices

- Reason abstractly and quantitatively.
- Construct viable arguments and critique the reasoning of others.

Vocabulary

- equals
- addition
- number sentences
- subtraction
- equation
- true
- false

Materials

- equation cards bag
- Equation Cards student page
- Make It True, Make It False student page
- counters for students to use to model the expressions

Task 20
Make It True, Make It False

Write true/false addition and subtraction equations

TASK

Make It True, Make It False

True or False?

Sadie and Sarai are playing a game together. They have a bag that contains equations written on cards. The object of their game is to work together to make as many true equations as they can. After that, they make false equations with their remaining pieces and then rewrite them to make them true.

TASK PREPARATION

- As first graders grow more comfortable using number sentences or equations, it is important to ensure that they understand the meaning of and appropriate use of the equal sign (=). One way to do this is through the use of true/false equations.

- Students will work in pairs for the majority of the task but will then share their equations with the rest of the class.

- Each pair of students will need a prepared set of equation cards, a Make It True, Make It False student page, and access to counters for those who want to model the expressions.

LAUNCH

1. Say, "As a class, let's start today by making a list of equations. For example, $2 + 1 = 3$ is an equation. Turn and talk with a partner and try to come up with as many other equations as you can."

2. After giving the class time to do so, call on students to generate a list of equations that you write on the board.

3. Create two columns "True" and "False." Say, "Now for each of these equations, we need to decide if they belong in the True column or the False column. If the equation is accurate, we will put it in the True column. If it not accurate, we will put it in the False column."

4. Say, "From our list of equations, work with your partner to decide if they belong in the True column or the False column." Ask students to *Turn and Learn* from a partner how they know that one of the equations on the board is true and how one of the equations is false. Provide counters for students to use to model the equations.

5. Have several students share their justifications. For example, for 2 + 1 = 3, a student might say, "I counted out 2 blocks and then added another one. So that's 3 blocks, and that's the same answer: 3."

6. Share the beginning of the task with students. Say, "You are going to work in teams of two with a baggie that has equation cards in it. Your goal today is to use these cards and create as many equations as possible that are true. Write those on the equation sheet. With your leftover cards, create equations that are false and put those on the equation sheet. Then rewrite the equations that are false and make them true."

FACILITATE

1. Distribute the baggies as well as the equation sheets to the students and have the student pairs complete the task. Also make sure that students have access to counting materials to help them prove whether the equation is true or false.

2. **Observe.** Monitor student progress with the task. Look for the following:

   » Are students able to write equations that are true?

   » Are the equations all of the same form? For students who generated many equations using the same format, you may want to emphasize that equations can be true in many different ways. Pair them up with a student who has written a true equation in a different form.

   » Are students using only one mathematical operation?

3. As you observe, interview selected student pairs. Consider asking the following:

   » Is this equation true or false?

   » How do you know?

   » What else can you tell me about what you did?

Note: Consider using the Observation or Interview (individual or small group) tools (see Appendix B).

> **STRENGTHS SPOTTING**
>
> Students may share that they feel more comfortable writing the same type of true equation over and over again. Celebrate the strength in the success of finding a solution that always works, while encouraging multiple strategies as a goal for all math problems. Leverage this strength by asking, "How can you use this true equation to find another?"

> **! PRODUCTIVE STRUGGLE**
>
> Students may not be familiar with the idea of justifying their mathematical decisions. True/false equations with access to counting materials provide a nice initial context for justification.

CLOSE: MAKE THE MATH VISIBLE

1. Once each student has finished their Make It True, Make It False page, have them each select and record an example as part of a Notice and Wonder Gallery Walk.

2. Bring the class back together to share their results.

3. Highlight the following:

» The meaning of the equal sign—that the expression on one side has the same value as the expression on the other side of the equal sign. Do this as a follow-up question to "How do you know if an equation is true?"

» Addition operations, subtraction operations, and numbers without operations can appear on both sides of true equations. Select at least one example where the right side of the equation is of the form c + d or c – d. Students are used to seeing equations of the form a + b = c and often believe that if an equation is not in that form, it is false. You want to have a discussion about examples where that is not the case. If there are no student examples of this form, offer 3 + 2 = 6 – 1 and 3 = 5 – 2.

STRENGTHS SPOTTING

Providing opportunities for students to make conjectures and then collaborate with peers as they try to prove or disprove them fosters development of strengths in Communication and in Reasoning and Proof—two critical components of Mathematical Proficiency (NRC, 2001).

TASK 20: MAKE IT TRUE, MAKE IT FALSE STUDENT PAGES

 To download printable resources for this task, visit **resources.corwin.com/ ClassroomReadyMath/K-1**

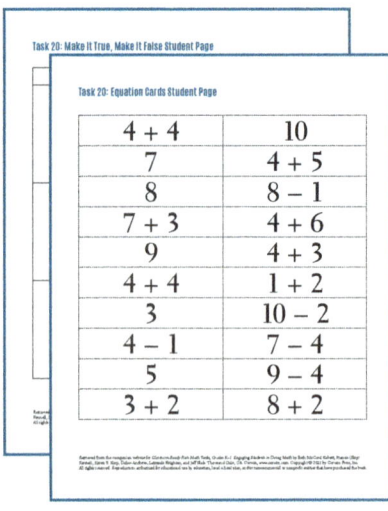

POST-TASK NOTES: REFLECTION & NEXT STEPS

Task 21
Splat and Split

Discover different decompositions of a counted collection

Mathematics Standard

- Decompose numbers less than or equal to 10 into pairs in more than one way, e.g., by using objects or drawings, and record each decomposition by a drawing or equation (e.g., 5 = 2 + 3 and 5 = 4 + 1).

TASK

Splat and Split

Figure 9.1 A Box Lid and a Paper Plate

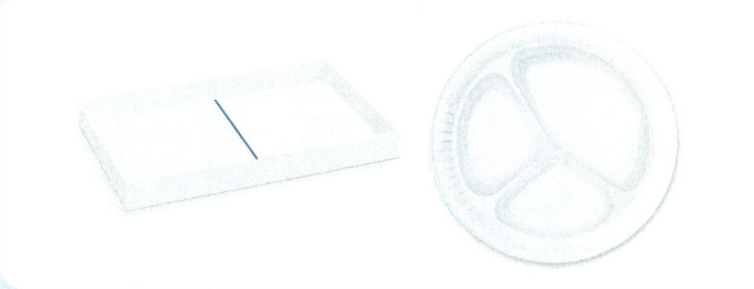

Source: Box lid by AzFree/iStock.com; paper plate by mastaka/iStock.com

Madeline and Hector found an old box lid in the basement. They decided to create a game. They drew a line across the middle of the lid and then took turns dropping handfuls of 10 counters over the lid. Madeline got points for every counter that landed on one side of the lid. Hector got points for every counter that landed on the other side of the lid. If they drop 10 counters each time, how many points might each person get?

Mathematical Practices

- Look for and make use of structure.
- Look for and express regularity in repeated reasoning.

Vocabulary

- number names from 0 to 10
- add
- equation
- plus sign (+)
- equal sign (=)

Materials

- box lids, trays, or paper plates (see Figure 9.1)
- collections of 1–20 items
- Splat and Split student page, one per student

TASK PREPARATION

- Prepare "drop zones" for each student pair by making a line across the center of a box lid, tray, or paper plate. Box lids will need to be large enough for the counting collection to spread out when dropped, and the raised edge will help keep collections contained.

- Prepare collections of countable items (counters, bears, counting cubes, etc.). Use items with flat surfaces so they will not roll.

- Decide whether all pairs will have the same total number in their collections or if different pairs will have different totals. For example, if the whole class has the same number of objects, then the focus will be on how many ways the collection can be distributed. If students are given different total amounts, the focus will be noticing general patterns in the relationships between addends and sums.

- If using different total amounts, consider how you will differentiate for student pairs.

LAUNCH

1. Organize students into heterogeneous pairs and distribute a collection of 1–20 objects to each pair. For example, some pairs will have 10 objects while other pairs will have 12.

2. Facilitate a *Pair-to-Pair Share* to allow students to talk about their collections. Be sure to partner pairs with distinct collections (bears and cubes or different color cubes) so it will be easy to avoid mixing them up. Have each pair decide who will drop the items first and who will record. Encourage the students to take turns dropping and recording both parts, recognizing that each part of the split adds to the total.

3. Ask students to share out some ways they figured out how many different combinations they could each have with another partner pair. Students should notice that their collections are not all the same size.

4. Distribute a student page to each student. Ask students to draw a math sketch to represent their collection.

5. **Hinge Question.** How many are in your collection? How do you know?

> **STRENGTHS SPOTTING**
>
> Highlight student pairs who are working well together by discussing their ideas and collaborating on decisions about who will drop and who will record.

> **ACCESS AND EQUITY**
>
> Students may not yet be proficient with making quick math sketches. Consider modeling how a circle or an "x" could be used to represent each item in the collection. Alternatively, students could be allowed to use dot markers to represent their collections on the student page. Provisions of this nature will ensure that students have time to move into the reasoning about different ways a collection may be decomposed, which is the central purpose of this task.

FACILITATE

1. Share the task story with the class.

2. Provide each pair with a box lid or tray that has been divided into two sections.

3. Ask, "What are some possible scores you think Madeline and Hector might get if they drop 10 counters (5 and 5, 4 and 6, etc.)? Why?"

4. Allow students a few moments to test out their ideas by dropping their collections in their lids, then have students share ideas with the class.

5. Provide time for each pair to try several rounds of the game. As they play, both students will record scores on their own student page.

6. **Interview.** As students play the game, circulate to each group. Ask:

   » How many (bears, counters, etc.) do you have in all?

   » How many landed on your side?

   » What do you notice about the scores you wrote down and the scores your partner wrote down?

   » Do you think there are any other ways the (bears/counters) could have landed?

   » Is it possible to have (8*) of the (bears/counters) land on your side?
   Suggest an amount greater than the total collection.

   » Is it possible to have none of the (bears/counters) land on your side? How would you record that?

Note: Consider using the Interview (individual or small group) tool (see Appendix B).

CLOSE: MAKE THE MATH VISIBLE

1. After students have had time to complete several rounds of the activity, facilitate another *Pair-to-Pair Share* so students can share their data. Be sure to partner up pairs with the same-sized collections.

2. As students share, **observe**. Pay attention to the patterns and relationships students notice.

   » Do students notice that the two scores always add up to the total amount in the collection?

   » Do students notice that when one score is increased, the other score is decreased?

   » Do students see the commutative relationship in the two partner recording sheets (see Figure 9.2)?

Figure 9.2 Sample Partner Splat and Split Student Pages

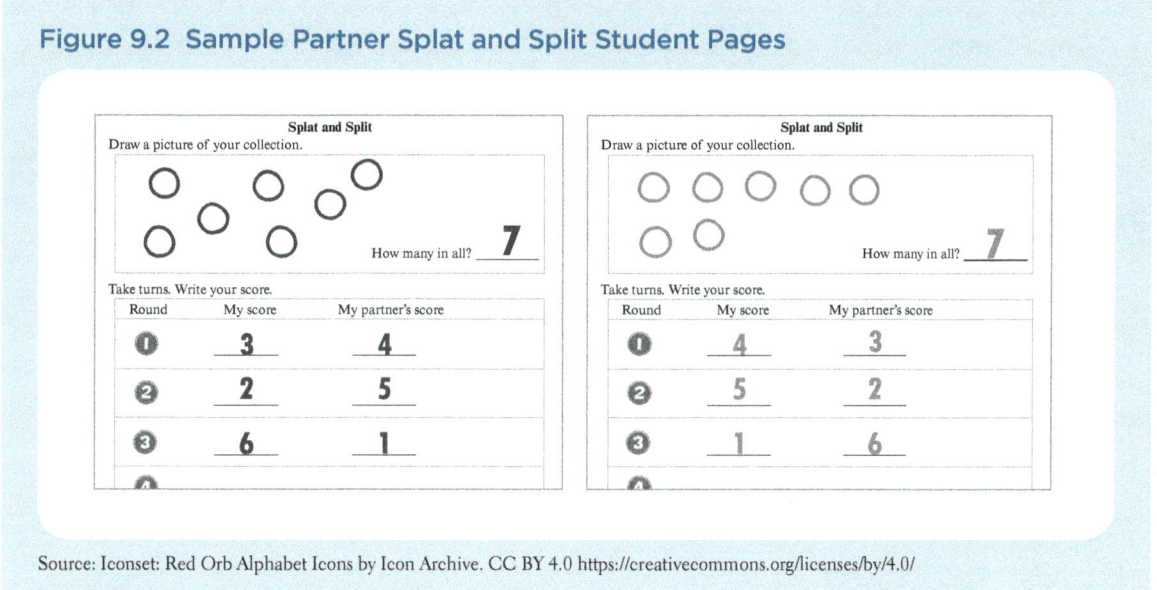

Source: Iconset: Red Orb Alphabet Icons by Icon Archive. CC BY 4.0 https://creativecommons.org/licenses/by/4.0/

3. Bring the whole class together and have selected quartets share some of the highlighted ideas heard during your observation.

 **PRODUCTIVE STRUGGLE**

Rather than providing rules or reframing student thinking, encourage students to use their drawing and physical collections to support their reasoning in their own words. Ask questions to support connections between the different representations: "How can you use your picture to show that 7 is equal to 3 plus 4?"

STRENGTHS SPOTTING

Students with a strong sense of mathematical agency will be wise consumers of mathematical ideas. Help students recognize the power in asking, "Does this make sense? Can it be proven?"

4. **Exit Task.** Say, "Some of you noticed that your scores added up to the total number of objects in your collection" (use student work to show an example like the one in Figure 9.3).

Figure 9.3 Student Work Example

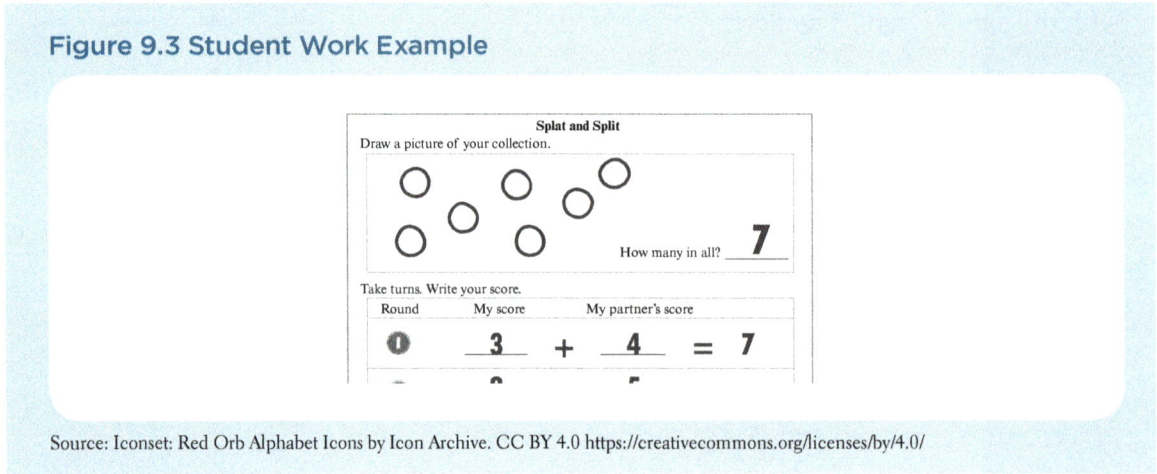

"Write equations to find the total for all of your scores. What do you notice?"

TASK 21: SPLAT AND SPLIT STUDENT PAGE

online resources — To download printable resources for this task, visit **resources.corwin.com/ ClassroomReadyMath/K-1**

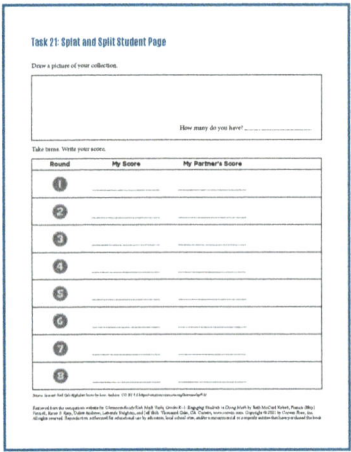

POST-TASK NOTES: REFLECTION & NEXT STEPS

Task 22
What's My Card?

Use subtraction as the inverse of addition

Mathematics Standard

- **Understand subtraction as an unknown-addend problem.** *For example, subtract 10 – 8 by finding the number that makes 10 when added to 8.*

Mathematical Practice

- Reason abstractly and quantitatively.

Vocabulary

- addition
- addend
- subtraction

Materials

- decks of playing cards
- What's My Card? student page, one per team

TASK

What's My Card?

Figure 9.4 Two Cards

Source: TokenPhoto/iStock.com

This card game is played with three players. Two players draw a card (see Figure 9.4) and put it on their forehead so that they can see each other's cards, but not their own. The third player announces the sum of the two cards. The two players then figure out the value of their card and explain their solution strategy to the other players. The third player records the card values as well as the equation that represents the solution strategy.

TASK PREPARATION

- Print out the What's My Card? recording sheet (see Figure 9.6) and assemble the number cards from a deck of playing cards, with enough decks so that the entire class can play in groups of three.

LAUNCH

1. Arrange the class for a whole-group discussion. Select two students to play the task game "What's My Card?"

2. Explain the directions to the students: "You will play this game in groups of three. Two players will each pick a card and put it on their forehead, so that the other players can see the card, but they cannot see their own card." Have two students each draw a card and put it on their forehead (see Figure 9.5). The ace is worth 1.

Figure 9.5 Students With Cards

Source: Cards by TokenPhoto/iStock.com; Student faces by Sudowoodo/iStock.com

3. Say, "The third player adds the two numbers using mental math and tells the two players who have cards on their forehead what the sum of the two cards is. In this case, the two cards are 3 and 4, so the sum is 7. Now, with that information, the two players with cards need to look at the other person's card and figure out what their own card is." Allow students to answer and explain.

Figure 9.6 Game Recording Sheet 1

What's My Card?

Card 1	Card 2	Sum	Equation 1	Equation 2
____	____	____	_____	_____
____	____	____	_____	_____

Figure 9.7 Game Recording Sheet 2

What's My Card?

Card 1	Card 2	Sum	Equation 1	Equation 2
4	3	7	3 + 4 = 7	4 + 3 = 7

4. Consider the example demonstrated in Figure 9.7.

   » Emilio said that he was told the total was 7 and he could see Tonisha's card was a 4. So Emilio started at 4 and counted on to get to 7. He counted on 3 numbers to get from 4 to 7, so he knew his card was a 3.

   » Tonisha said that she was told the sum was 7 and Emilio's card was 3, so she counted on 4 more to equal the sum of 7. The missing addend was 4.

   » The third player, Isaac, now records what the two cards were as well as the sum, writing equations for the explanation. Isaac writes on the recording sheet 3 + 4 = 7, 4 + 3 = 7, 7 – 4 = ?, and 7 – 3 = ?

5. "Is everyone ready to play?"

FACILITATE

1. Organize students into heterogeneous groups of three.

2. Distribute a set of cards and a recording sheet to each group. Counters should also be made available for students if they would like to use them.

3. Ask students to start playing What's My Card?, changing roles with each two new cards drawn.

4. **Observe.** As students work, circulate through the groups and note how students are explaining their thinking in arriving at their card value.

5. **Interview.** Ask each group:

   » What strategies have you used so far? Which of these strategies was most helpful?

   » How can you use addition to figure out your card value?

   » How might you use subtraction?

   » How are the two equations that you have written the same? How are they different?

Note: Consider using the Observation and Interview tools to monitor student understanding (see Appendix B).

ACCESS AND EQUITY

Heterogeneous groups allow for a wider range of strategies. Grouping students in this way provides them with opportunities to learn from and support one another.

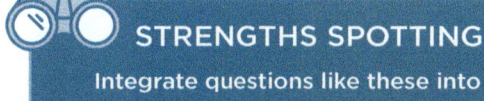

STRENGTHS SPOTTING

Integrate questions like these into your regular routine to help your students share their mathematical strengths. Be sure to note the students' strategies as mathematical strengths.

CLOSE: MAKE THE MATH VISIBLE

1. After students have had multiple turns playing the game in different roles, bring the class together and have the groups share observations.

2. Select and Sequence small groups to share their thinking based on some of the interview questions asked during the Facilitate phase of the task. Highlight the following ideas:

   » Addition and subtraction strategies can both be viable ways of solving the same problem.

   » When a student uses an addition/counting-up strategy, have them also generate the subtraction strategy.

   » Similarly, have a student who uses a subtraction strategy use the appropriate addition/counting-up strategy.

3. **Exit Task.** If you are playing What's My Card? and you know the sum is 10 and the card you can see is a 6, what is your card? Show me how to solve this problem in two ways: by counting up and by using subtraction.

TASK 22: WHAT'S MY CARD? STUDENT PAGE

POST-TASK NOTES: REFLECTION & NEXT STEPS

_____

_____

_____

_____

_____

_____

_____

_____

_____

Task 23

Lots of Equations

Determine the unknown in addition or subtraction equations

TASK

Lots of Equations

Figure 9.8 Pharrell and Zaina's Cards

Source: TokenPhoto/iStock.com

Pharrell and Zaina are playing a card game. Pharrell flips over two cards to reveal two numbers: 5 and 8 (see Figure 9.8). Now Pharrell and Zaina write as many addition and subtraction equations as they can for those two numbers. Pharrell writes $5 + \boxed{3} = 8$ and $5 + 8 = \boxed{13}$. Zaina uses subtraction and writes $8 - 5 = \boxed{3}$ and $8 - \boxed{3} = 5$. Can you think of other equations?

TASK PREPARATION

- The Launch will be with the whole class, with the examples involving student pairs.

- The task activity will involve students working in small groups.

LAUNCH

1. Bring all students to whole-group discussion. "We are going to learn a new card game today! This game is called Lots of Equations! Let's learn how to play. First, I need a volunteer to help me lay out 16 cards." Select a student to arrange 16 cards in four rows of four.

2. Say, "Now flip over any two cards and read out the number on each of the two cards."

3. Say, "Turn and talk with your neighbor. What are some different equations that you could write using these two numbers? Put a box around a number that is unknown in each

Mathematics Standard

- Determine the unknown whole number in an addition or subtraction equation relating three whole numbers. For example, determine the unknown number that makes the equation true in each of the equations $8 + ? = 11$, $5 = \underline{\quad} - 3$, $6 + 6 = \underline{\quad}$.

Mathematical Practices

- Reason abstractly and quantitatively.

- Look for and make use of structure.

Vocabulary

- addition

- subtraction

- equations

Materials

- Number Cards student pages (1–13), two sets for each group

- Lots of Equations student page (one per group), cut in two for future sorting of equations

- sticky notes, one set per group of 3

- counters or connecting cubes (provide access to all students)

- tape to display the posters

- whiteboards for the teams (optional)

of the equations you discuss. Include addition and subtraction equations *and* equations where the change is unknown."

4. Have the student pairs generate equations and then write them on the whiteboard.

5. Say, "Record your equations on the recording sheet." Check for understanding of the game directions. "Turn and talk with your partner to make sure everyone understands the game. Now you try it with your small groups."

FACILITATE

1. Distribute a deck of number cards and a recording sheet to each small group.

2. Ask students to start playing Lots of Equations, taking turns with each two new cards.

3. **Observe.** As students work, circulate through the groups and note the types of equations that students are writing. Look for the following: Do their equations use addition or subtraction or both? Are they varying the location of the unknown (change, start, result)?

4. As you observe, also interview the small groups. Some possible interview questions include the following:

   » Show me some different equations that your group wrote.

   » How are your equations similar? How are they different?

   » Tell me which one of your equations is a Result Unknown, a Change Unknown, or a Start Unknown?

PRODUCTIVE STRUGGLE

It's important to resist the temptation to step in too soon to provide guidance on an open-ended task. If students need assistance, ask them to think about strategies that they know for adding and subtracting numbers.

Note: Consider using the Observation and Interview (small group) tools here as you look for and note student equations (see Appendix B).

CLOSE: MAKE THE MATH VISIBLE

1. Allow students to complete a Gallery Walk to view all of the completed recording sheets. As the students participate in the Gallery Walk, ask:

   » Which equations do your classmates have that you do not?

   » Which equations do you have that no one else does?

2. Bring students back to whole-group discussion. If possible, find two numbers that multiple groups used and allow students to present and discuss all equations that were created using those two numbers.

3. Select and Sequence groups to share their thinking based on some of the interview questions asked during the Facilitate phase of the task.

STRENGTHS SPOTTING

Throughout this task, there have been opportunities for students to help solve problems with a partner or in small groups. These opportunities give you ways to spot strong problem-solving behaviors in your students.

4. Highlight the following ideas:

» The equations could have used addition or subtraction. For example, with the numbers 8 and 3, 8 + 3 = ☐11☐, 3 + 8 = ☐11☐, and 8 – 3 = ☐5☐ could all be used.

» The location of the unknown could be at the start, change, or result. Using the numbers 3 and 8, ☐5☐ + 3 = 8, 3 + ☐5☐ = 8, ☐11☐ – 3 = 8, ☐11☐ – 8 = 3, and 8 – ☐5☐ = 3 are all examples of viable equations.

TASK 23: LOTS OF EQUATIONS STUDENT PAGES

 To download printable resources for this task, visit **resources.corwin.com/ ClassroomReadyMath/K-1**

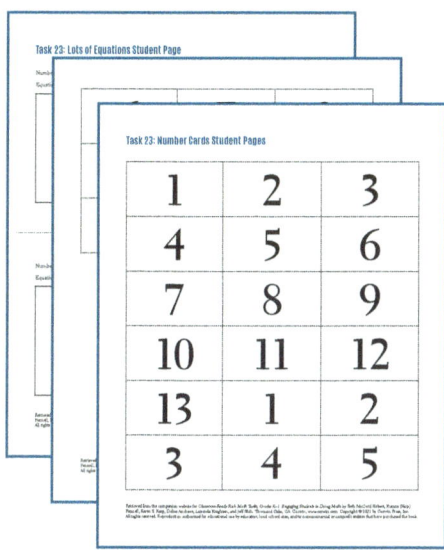

POST-TASK NOTES: REFLECTION & NEXT STEPS

Mathematics Standard

- For any number from 1 to 9, find the number that makes 10 when added to the given number, e.g., by using objects or drawings, and record the answer with a drawing or equation.

Mathematical Practices

- Look for and make use of structure.
- Look for and express regularity in repeated reasoning.

Vocabulary

- bowling
- bowling pin
- equation

Materials

- At the Bowling Alley Bowling Pin student pages (bowling pin cards and individual bowling pin counter cards to cut out for counters), one set for each group
- counters to represent pins (if you don't cut out pins)
- whiteboards or scratch paper
- sentence strips

Task 24

At the Bowling Alley

Find the missing addend to make 10

TASK

At the Bowling Alley

Oscar was bowling with his family. In the game of bowling, each player gets two tries to knock down 10 bowling pins using a bowling ball. After Oscar's first try, there are 4 pins left (see Figure 9.9).

Figure 9.9 Oscar's Pins

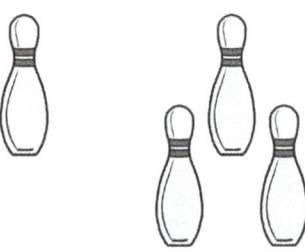

Source: pixabay.com

How many pins did Oscar knock down?

TASK PREPARATION

- Consider inviting a bowling team member from your local community or a student's family member who bowls to come and show the class what a bowling pin and bowling ball look like, and answer questions about bowling.

LAUNCH

1. Facilitate a Notice and Wonder using a quick video of someone bowling. For example, watch minutes 20:58–21:06 of the 2019 NCAA Women's Bowling Championship at the following link: youtube.com/watch?v=TLddeFyYHjo (watch without audio). This clip shows someone knocking down all but two pins in one turn.

2. Record students' observations and questions on the board.

3. Explain to students that in the game of bowling, each player gets two tries to knock down 10 bowling pins using a bowling ball. Show a picture of how bowling pins are arranged (see Figure 9.10) and ask students to count the pins.

Figure 9.10 How Bowling Pins Are Arranged

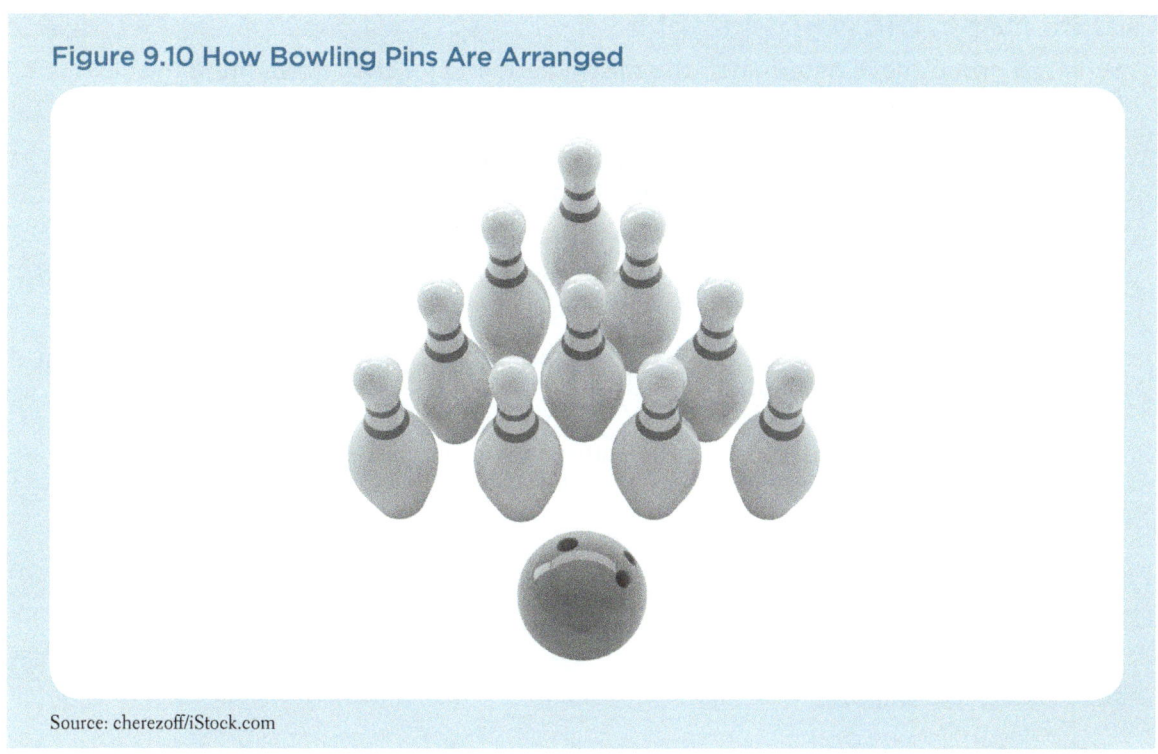

Source: cherezoff/iStock.com

FACILITATE

1. Present the task to students.

2. Organize students into heterogeneous groups of two or three. Provide each group with a different bowling pin card and access to a supply of individual bowling pin cutouts.

3. Tell students that the image shows how many bowling pins are still standing after a bowler's first turn. Challenge students to figure out how many pins the bowler has already knocked down.

4. As students work, circulate through the class to observe. Take note of the methods students use to determine the missing number of pins. Some approaches to watch for:

   » Are students counting on from the number of pins on their card to reach 10? If so, how do they keep track of how many more they count on?

   » Are students drawing sketches to fill up to 10?

   » Do students recognize a combination that adds to 10 and "just know" how many are missing? If so, can they prove their answer is correct?

Note: Consider using the Observation tool to record the student methods for determining the missing number of pins (see Appendix B).

5. As students finish, ask them to write an equation on a sentence strip to show that their answer added to the number of pins shown on their card will equal 10 pins. These can be used to support student sharing in Close, step 3.

 **PRODUCTIVE STRUGGLE**

To provide space for productive struggle to happen, avoid directing students to a specific strategy. Allow students to explore to discover how they might determine the number of pins knocked down.

 **STRENGTHS SPOTTING**

Seeking to understand a student's reasoning for whatever organizational structure they've chosen (even if initial choices are less efficient) signals that their ideas are valued. Purposeful questioning and opportunities to explore connections between representations will support growth from a place of strength.

CLOSE: MAKE THE MATH VISIBLE

1. When all groups have determined the missing addend for their card, bring the class back together.

2. Allow each group to show their card and tell how many pins are left, how many were already knocked down, and the equation they wrote on a sentence strip.

3. As each group shares, display the sentence strip equation (from Facilitate, step 5) or have one of the group members write an equation on the board or on a class poster to show that the total amount is equal to 10 pins (left standing up + knocked down = 10).

4. When all the equations are recorded, have students *Turn and Learn* from a partner. Ask, "What are you noticing about all of our equations?"

5. **Exit Task.** When it was her turn, Oscar's aunt knocked down 7 pins on her first try. How many pins are still standing? Show how you know.

TASK 24: AT THE BOWLING ALLEY STUDENT PAGES

online resources ↘ To download printable resources for this task, visit **resources.corwin.com/ ClassroomReadyMath/K-1**

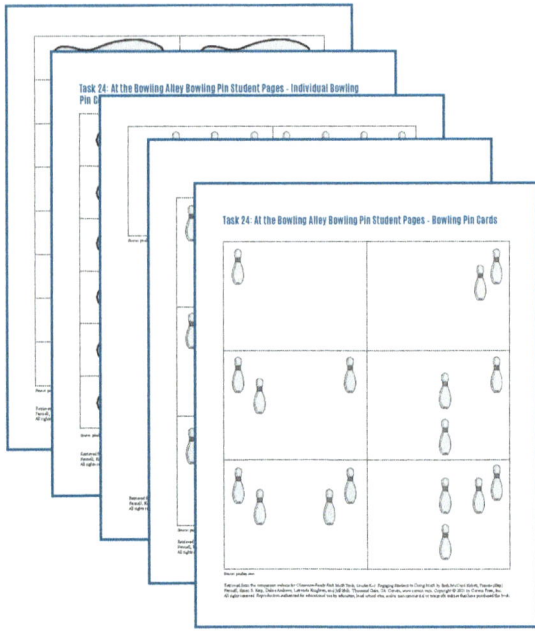

POST-TASK NOTES: REFLECTION & NEXT STEPS

Task 25
Keep Making Tens!

Make tens

TASK

Keep Making Tens!

Figure 9.11 A Cooperative Card Game

Source: TokenPhoto/iStock.com

This cooperative card game is played with two players. The players place 12 number cards face up in an array between the two of them. Each player takes turns selecting two cards that sum to 10 and keeps the match in a pile in front of them (see Figure 9.11). After the appropriate equation is written, two cards are replaced into the array from the deck and the other player takes their turn. The game ends when there are no more cards left in the deck and there are no combinations left.

TASK PREPARATION

• This task provides an opportunity for students to play a cooperative game that provides them with opportunities to make combinations that equal 10 and explain their thinking.

• Assemble two sets of number cards into an 18-card deck for each pair of students.

• Also, have counting cubes available for those students who need them.

Mathematics Standard

• Add and subtract within 20, demonstrating fluency for addition and subtraction within 10. Use strategies such as counting on; making ten (e.g., 8 + 6 = 8 + 2 + 4 = 10 + 4 = 14); decomposing a number leading to a ten (e.g., 13 – 4 = 13 – 3 – 1 = 10 – 1 = 9); using the relationship between addition and subtraction (e.g., knowing that 8 + 4 = 12, one knows 12 – 8 = 4); and creating equivalent but easier or known sums (e.g., adding 6 + 7 by creating the known equivalent 6 + 6 + 1 = 12 + 1 = 13).

Mathematical Practice

• Construct viable arguments and critique the reasoning of others.

Vocabulary

• numbers 1 through 10
• equals
• equal sign
• sum

Materials

• Number Cards student page (1–9), two decks for each team
• Keep Making Tens! student page
• counting cubes

LAUNCH

1. Bring the class together for a whole-group discussion. Select two of your students to play the card game Keep Making Tens.

2. Say,

   » "You will play this game in groups of two. Two players, Daryl and Natalia, will set out 12 cards in rows and columns so that you can see the numbers." Have one of the students arrange 12 cards (out of the 18 total cards from the two decks) in three rows of four so that both students can see the 12 cards.

   » "Daryl looks at the cards and picks two of them that add up to 10. He writes the equation for those two cards on the recording sheet and explains how he got the sum to Natalia."

   » "Natalia asks Daryl questions until she understands how he got his sum. Then she takes two cards from the remaining deck to replace the two that Daryl just used."

   » "Now it's Natalia's turn to pick two cards that sum to 10. She writes the equation on her recording sheet and explains her thinking to Daryl, and Daryl asks her questions until he understands her thinking."

   » "They keep playing until there are no cards in the deck left to play and no more combinations of 10."

3. Ask the students to *Turn and Talk*. Ask, "Who won? What do you notice about their equations?"

> **! PRODUCTIVE STRUGGLE**
>
> Some students might still need to use the counting cubes to help them determine the sum. Using the numbers in a Add to, Result Unknown word problem may help those students who would benefit from the context and action of a word problem.

> **STRENGTHS SPOTTING**
>
> Just as explaining strategies and ideas are students' strengths, so too are the processes and practices of having students listen to each other's ideas. Highlight students who are demonstrating listening behaviors and consider naming the listening behaviors you are noting.

FACILITATE

1. Organize students into heterogeneous pairs to play the game.

2. Distribute two sets of number cards (1–9 each for a total of 18 cards in the deck) and a recording sheet to each pair. Have them organize the cards face up in a 3 × 4 array.

3. Ask students to start playing Keep Making Tens, as Daryl and Natalia did, taking turns picking card pairs, writing each new equation, and explaining their thinking.

4. **Observe.** As students play, circulate through the pairs and note how students are explaining their thinking in arriving at their sum.

5. **Interview.** Ask each pair:

   » What strategies have you used so far?

   » Show me an example of your partner's solution strategy.

   » How are the equations that you have written the same? How are they different?

Note: Consider using the Observation, Interview (small group), and Show Me tools for both monitoring progress (Observation) of the pairs and recording responses (Interview and Show Me) (see Appendix B).

CLOSE: MAKE THE MATH VISIBLE

1. After students have had multiple turns (at least three or four) playing the game, bring the class together and have the pairs share observations. As they do, briefly review their recording sheets.

2. Select and Sequence pairs to share their thinking based on some of the Interview and Show Me responses from the Facilitate portion of the task lesson. Highlight that there are multiple ways to add the number cards, and each of these ways is valuable.

   » Some students use the counting cubes to represent both numbers and count them all.

   » Some students count on from one of the numbers.

   » Some students recall the sums to 10.

TASK 25: KEEP MAKING TENS! STUDENT PAGES

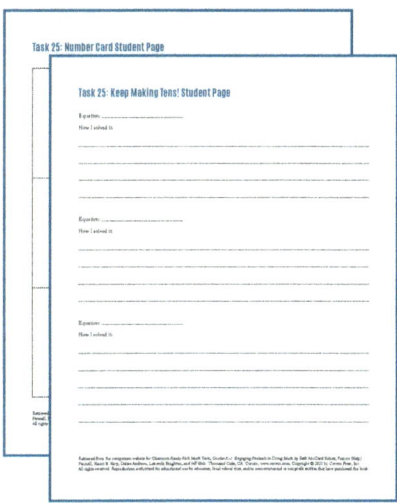

To download printable resources for this task, visit **resources.corwin.com/ ClassroomReadyMath/K-1**

POST-TASK NOTES: REFLECTION & NEXT STEPS

Number and Operations in Base Ten

Place Value

GETTING STARTED

TASK 26: KINDERGARTEN: BUTTONS IN A POCKET

Compose and decompose numbers from 11 to 19 into ten ones and some further ones, e.g., by using objects or drawings, and record each composition or decomposition by a drawing or equation (e.g., 18 = 10 + 8); understand that these numbers are composed of ten ones and one, two, three, four, five, six, seven, eight, or nine ones.

TASK 27: GRADE 1: WHAT'S MISSING?

Understand that the two digits of a two-digit number represent amounts of tens and ones. Understand the following as special cases:

a. 10 can be thought of as a bundle of ten ones—called a "ten."

b. The numbers from 11 to 19 are composed of a ten and one, two, three, four, five, six, seven, eight, or nine ones.

c. The numbers 10, 20, 30, 40, 50, 60, 70, 80, 90 refer to one, two, three, four, five, six, seven, eight, or nine tens (and 0 ones).

TASK 28: KINDERGARTEN: CUPCAKE DECORATING

Compose and decompose numbers from 11 to 19 into ten ones and some further ones, e.g., by using objects or drawings, and record each composition or decomposition by a drawing or equation (e.g., 18 = 10 + 8); understand that these numbers are composed of ten ones and one, two, three, four, five, six, seven, eight, or nine ones.

TASK 29: GRADE 1: DIGIT COMPARE

Understand that the two digits of a two-digit number represent amounts of tens and ones.

TASK 30: GRADE 1: COIN COLLECTIONS

Compare two two-digit numbers based on meanings of the tens and ones digits, recording the results of comparisons with the symbols >, =, and <.

TASK 31: GRADE 1: NUMBER DISAGREEMENT

Understand that the two digits of a two-digit number represent amounts of tens and ones.

Anticipating Student Thinking: As early learners begin their mathematical journey, often informal experiences involving counting and related conceptual understandings are foundational, as are such experiences with place value. At these grade levels, that means tens and ones. Consider that place value is centrally connected within a continuum that begins with counting and extends counting to understanding numbers greater than nine, which presents the need for a direct focus on place value. The tasks presented in this chapter focus directly on the importance of place value, in particular tens and ones. As you consider the implementation of the chapter's tasks, make sure to provide the time necessary for your students to truly engage with the representation materials (e.g., base ten blocks, ten frames, counters, pictures) presented in the implementation portion of the task lesson (Launch, Facilitate, and Close) and that connect directly to early place value understandings.

> **THINK ABOUT IT**
>
> An important hallmark of any *doing-math task* is student engagement. As you implement the chapter's tasks, regularly use Observation as a formative assessment technique to monitor student engagement, making sure that all students have access to the task and are fully engaged in the task's activities.

Task 26
Buttons in a Pocket

Represent teen numbers as 10 and some more ones

TASK

Buttons in a Pocket

Figure 10.1 Some Buttons

Source: pixabay.com

Maria has 16 buttons, 9 in her right pocket and 7 in her left pocket. Jamar has 15 buttons, 10 in his right pocket and 5 in his left pocket. Maria said they could fill ten frames with their buttons to compare their amounts. Draw on the little ten frames how many buttons each of them has. How many tens and some more does Maria have? How many tens and some more does Jamar have? You may use counters and ten frames to help you solve this problem.

TASK PREPARATION

- Students will work in pairs on this task.

LAUNCH

1. Tell students the beginning of the story: "Maria has 16 buttons, 9 in her right pocket and 7

ALTERNATE LEARNING ENVIRONMENT

If counters and ten frames are unavailable, consider using a digital resource such as the Number Frames app from the Math Learning Center (https://apps .mathlearningcenter.org/ number-frames/).

Mathematics Standard

- Compose and decompose numbers from 11 to 19 into ten ones and some further ones, e.g., by using objects or drawings, and record each composition or decomposition by a drawing or equation (e.g., 18 = 10 + 8); understand that these numbers are composed of ten ones and one, two, three, four, five, six, seven, eight, or nine ones.

Mathematical Practices

- Reason abstractly and quantitatively.
- Look for and make use of structure.

Vocabulary

- ten
- ones
- some more
- ten frames

Materials

- 40 buttons (see Figure 10.1) or counters in a baggie
- paper
- markers or crayons
- little ten frames, one for each student
- Buttons in a Pocket student page, two per student

in her left pocket. Jamar has 15 buttons, 10 in his right pocket and 5 in his left pocket." Ask, "What do you notice? What do you wonder?" Allow students 1 minute of independent think time. Then ask students to *Turn and Talk* to their partner and share their thinking about how they can compare the number of buttons. Allow time for several student pairs to share their discussions with the larger group. Record the students' ideas.

ACCESS AND EQUITY

Providing students with individual think time gives each student an opportunity to make their own observations before discussing with a partner. *Turn and Talk* allows students to verbalize their observations with a partner before sharing with the entire class and helps to build student confidence.

Notice	Wonder

2. Distribute the work mats and counters to student pairs. Allow time for the students to model the number of buttons in Maria's and Jamar's pockets using counters. Ask, "What do you notice? What do you wonder?"

3. **Observe** students as they model the number of buttons in Maria's and Jamar's pockets, making note of the strategies they use to show the quantities.

4. Tell students the next part of the story: "Maria said they could fill ten frames with counters to compare their amounts." Ask students, "How could you use ten frames and counters to compare Maria's buttons to Jamar's buttons? Turn and talk to your partner to share your thinking."

5. **Interview.** Visit student pairs and ask, "How do you know if Maria or Jamar has more buttons?" Record student ideas.

Note: Consider using the Observation and Interview (individual or small group) tools to monitor student understanding (see Appendix B).

FACILITATE

1. Distribute paper, markers or crayons, ten frames, and counters to student pairs.

2. **Interview.** Visit student pairs and ask, "Maria has 16 buttons. What do you know about the number 16?" Allow time for students to respond.

3. **Show Me.** Say, "Use the recording sheet of little ten frames and show me how many buttons Maria has as a number of tens and some ones left over. How many tens and how many more buttons does Maria have? You may use counters and ten frames to help you solve this problem." Remind students of their ideas about how to use ten frames and counters to compare Maria's buttons to Jamar's buttons. Say, "Turn and talk to your partner to explain how you know."

4. **Interview.** Visit student pairs and ask, "Jamar has 15 buttons. How can you describe 15 using the tens and leftover ones?" Allow time for students to respond.

5. Say, "Draw and show me how many buttons/counters Jamar has as ten ones and some ones. How many tens and how many more buttons/counters does Jamar have? Turn and talk to your partner to explain how you know. What if Maria moves some of her buttons from the left pocket to the right pocket? Will her number of tens and some more ones change?"

Note: Consider using the Interview (individual or small group) and Show Me tools to monitor student understanding (see Appendix B).

CLOSE: MAKE THE MATH VISIBLE

1. Bring the students together for a whole-group discussion to share their drawings.

2. Ask students to describe their strategies and solutions by telling how many counters Maria and Jamar have as a number of tens and some more ones. Student responses should include the following:

   » Maria has 1 ten and 6 more ones.

   » Jamar has 1 ten and 5 more ones.

TASK 26: BUTTONS IN A POCKET STUDENT PAGE

 To download printable resources for this task, visit **resources.corwin.com/ ClassroomReadyMath/K-1**

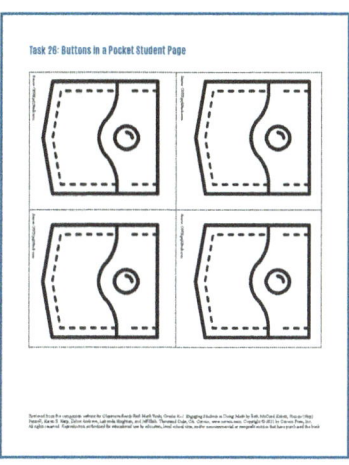

POST-TASK NOTES: REFLECTION & NEXT STEPS

Mathematics Standard(s)

- Understand that the two digits of a two-digit number represent amounts of tens and ones. Understand the following as special cases:

  a. 10 can be thought of as a bundle of ten ones—called a "ten."

  b. The numbers from 11 to 19 are composed of a ten and one, two, three, four, five, six, seven, eight, or nine ones.

  c. The numbers 10, 20, 30, 40, 50, 60, 70, 80, 90 refer to one, two, three, four, five, six, seven, eight, or nine tens (and 0 ones).

Mathematical Practices

- Attend to precision.
- Reason abstractly and quantitatively.

Vocabulary

- digit
- decompose
- compare
- tens
- ones

Materials

- open number line
- base ten materials
- What's Missing? Number Line student page

Task 27
What's Missing?

Using place value to determine the missing number on a number line

TASK

What's Missing?

What could the missing number on the number line (Figure 10.2) be? Explain why you picked your answers.

Figure 10.2 Number Lines With Missing Endpoints

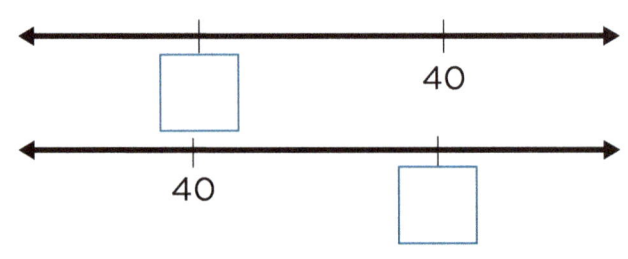

TASK PREPARATION

- Consider students' familiarity with a number line. If students have previously used number lines, begin the Launch with step 2.

LAUNCH

1. Ask students to respond to questions about a classroom-sized number line from 0–20, with the following numbers missing: 7, 13, and 19.

   » How many numbers are missing?

   » Which are the smallest and largest missing numbers?

   » What other number is missing?

2. Bring students together as a whole group and project an example of an open number line (Figure 10.3), using a number line from 20–50, to the entire class.

Figure 10.3 A Number Line With Endpoints Labeled

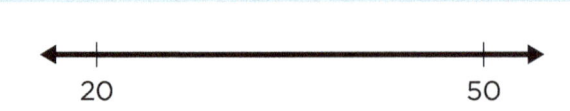

3. Ask, "What do you notice about this open number line?" Elicit from the students that the number line's endpoint numbers are 20 and 50.

4. Ask, "Now what would happen if a person comes by and erases the 20 on our number line?" (See Figure 10.4.)

Figure 10.4 A Number Line With One Endpoint Erased

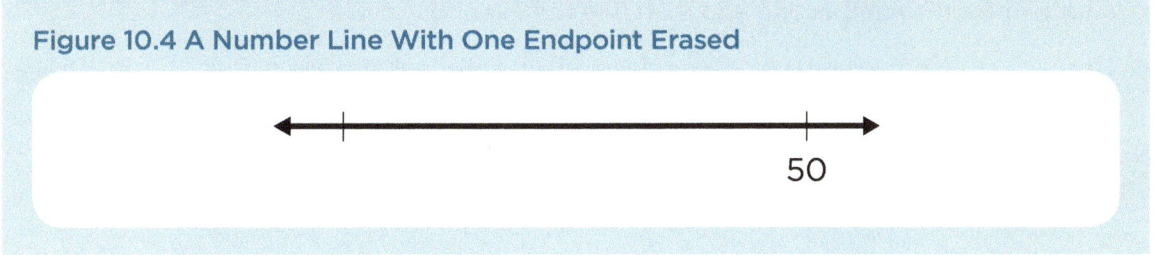

5. Elicit from the students that they wouldn't know what the end point on the number line is.

6. Say, "Now take a look at this number line (Figure 10.5). Pedro walks in and sees this number line and places 43 in the box."

Figure 10.5 A Number Line With One Endpoint Labeled

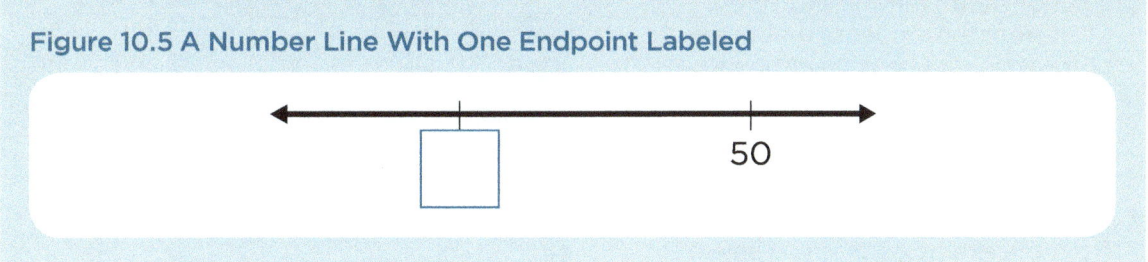

7. Ask, "Could Pedro be right? Turn and talk with a partner to explain how Pedro could be correct. What could Pedro's starting number be on the number line?" Elicit from the students what Pedro might be thinking.

8. Say, "What I would like you to think about today with a partner is, what number could go in that box?" Point at the 50, then ask, "How many tens and ones are in this number? Could more than one number go in that box? How do you know? I want you to be thinking about what you know about tens and ones. How many tens are in the number you want to put in the box?"

FACILITATE

1. Place students in pairs with their open number line tasks.

2. **Observe** as students make sense of the numbers they are placing on the number line. **Interview** pairs: "What number could go in the box? Why could that number go in the box? How many tens are in that number? If you put that number in the box, what numbers would go on the endpoints to make that number make sense?"

STRENGTHS SPOTTING

Students who demonstrate strengths in reasoning and proof often notice patterns and structures in word problems. Highlight, name, and celebrate these strengths when students exhibit them.

! PRODUCTIVE STRUGGLE

Mathematics tasks that include more than one possible correct answer could be new to students. These types of tasks broaden students' thinking so that they come to learn that mathematics problems don't always just have one answer.

3. Ask, "Could more than one number go in the box? How do you know? Is there more than one possible answer?"

4. Have students explain their ideas to their partners. Ask, "Did you get the same answer? Why or why not?"

5. If students could benefit from additional experiences, ask students to determine an endpoint for the number line and then find the missing number.

6. **Observe** students to see their strategies and note student strategies that you would like to share in the Close.

Note: Consider using the Observation and Interview tools to monitor student understanding (see Appendix B).

CLOSE: MAKE THE MATH VISIBLE

1. Bring students back together as a whole group. Ask the students to post their solutions on chart paper or a board for everyone to see. Ask, "What do you notice? Turn and talk with a partner about what you see."

2. Say, "I see that we have many different solutions. Can they all be correct?" Elicit from the students that many numbers could be correct because it depends on how they were thinking about the number values marked on the number line.

STRENGTHS SPOTTING

Recognize students who use multiple representations to demonstrate their understanding, particularly when they connect representations to procedures. Specifically call attention to how the students use the number line rather than using general praise.

3. Ask students to share their thinking from the examples that you selected for wider group exposure in the Facilitate session.

4. **Hinge Question.** Could the missing number in the second example be a one-digit number? Could the number be more than 50? Why or why not?

TASK 27: WHAT'S MISSING? NUMBER LINE STUDENT PAGE

 To download printable resources for this task, visit **resources.corwin.com/ ClassroomReadyMath/K-1**

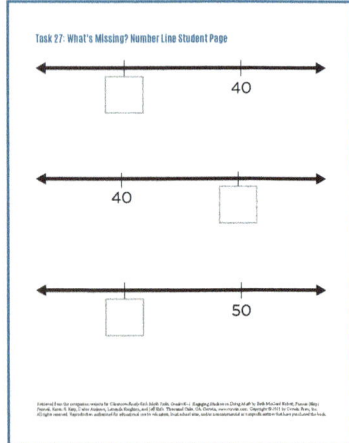

POST-TASK NOTES: REFLECTION & NEXT STEPS

_____

_____

_____

_____

_____

_____

_____

_____

_____

_____

_____

_____

_____

Task 28

Cupcake Decorating

Represent the sum of two one-digit numbers as a ten and some more ones

TASK

Cupcake Decorating

Figure 10.6 Brianna and Brian's Cupcakes

Source: pixabay.com

The twins Brianna and Brian are decorating cupcakes for their grandmother's birthday party (Figure 10.6). Brianna decorated 6 cupcakes and Brian decorated 8 cupcakes. If they put their cupcakes together, how many cupcakes will they have? Use ten frames to show how many cupcakes they have as groups of ten ones and some more ones. You may use counters to help solve this problem.

TASK PREPARATION

- Students will work in pairs on this task.

LAUNCH

1. Say, "The twins Brianna and Brian are decorating cupcakes for their grandmother's birthday party. Brianna decorated 6 cupcakes and Brian decorated 8 cupcakes."

ALTERNATE LEARNING ENVIRONMENT

If ten frames and counters aren't available, consider using a digital resource such as the ten frames virtual manipulative from Didax: www.didax.com/apps/ten-frame/

2. Distribute the ten frames and counters to student pairs. Say, "Partner A, use your counters to show me how many cupcakes Brianna decorated, and Partner B, use your counters to show me how many cupcakes Brian decorated."

3. Allow time for the students to model the cupcake amounts on their ten frames using counters. Ask, "What do you notice? What do you wonder?"

4. **Observe** students as they model the cupcake amounts.

Note: Consider using the Observation tool to record student understanding (see Appendix B).

FACILITATE

1. Distribute paper, markers or crayons, ten frames, and counters to student pairs.

2. Ask, "If Brian and Brianna put their cupcakes together, how many cupcakes will they have? Use ten frames to show how many cupcakes they have as groups of ten ones and some more ones. You may use counters and ten frames to help you solve this problem." **Observe** students to see how they organized their counters to count them. Did student pairs use the ten frames? If so, how did they use the ten frame to find their answers? Next, **observe** students to see what they do with the extra 4 counters after they fill one ten frame. (Students could also use portion cups to group objects by ten.) Note the students' strategies.

3. Distribute the Cupcake Decorating recording sheet and have students work together to draw cards and determine the number of tens and some more ones for each card. Observe students as they work together. Note the strategies they use.

4. Assign each pair of students to work with another pair of students to create a group of four for a *Pair-to-Pair Share*. Say, "Meet with your new friends and share your work. Look at the other pairs' work. What do you notice? Did they solve the problem the same way you did? Did they solve it differently?"

> ## ! PRODUCTIVE STRUGGLE
>
> To encourage sense making in students, allow time for students to make sense of the task and use problem-solving strategies to determine how to approach the task.

> ## ACCESS AND EQUITY
>
> Allowing students to self-select strategies and tools may provide access for students who have unique approaches to the problem. Purposeful questioning will facilitate connections between and among students' thinking.

CLOSE: MAKE THE MATH VISIBLE

1. Bring students together as a group. Say, "The twins put the cupcakes in a box like this one that holds 10 cupcakes. What happens when they fill the cupcake box?" (Possible student responses: "They have some cupcakes left over." "There are four cupcakes that don't fit in the box.") Ask, "How would you describe how many cupcakes they have using the language tens and some more ones?"

2. Ask, "How many cupcakes do the twins have for the party? Explain how you know." Allow time for students to respond. Student pairs should share that Brianna and Brian have 14 cupcakes if they put their cupcakes together. Student responses should include descriptions of the cupcakes as 1 ten and 4 more ones or 1 box of ten and 4 more cupcakes.

3. Ask student pairs to show how they solved the task.

4. **Hinge Question.** How would you describe 14 cupcakes as tens and ones? Show how you know.

5. **Exit Task.** Ella invited 9 girls and 7 boys to her birthday party. Draw a picture to show me how many children she invited to her birthday party. Show your answer in tens and ones.

TASK 28: CUPCAKE DECORATING STUDENT PAGES

 To download printable resources for this task, visit **resources.corwin.com/ ClassroomReadyMath/K-1**

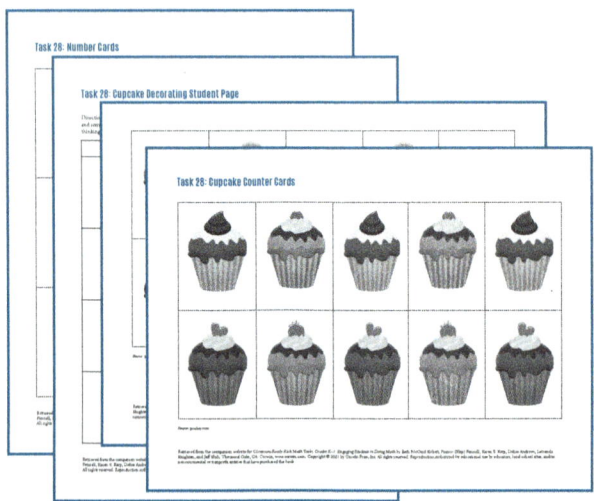

POST-TASK NOTES: REFLECTION & NEXT STEPS

Task 29

Digit Compare

Understand the meaning of each digit in a two-digit number

TASK

Digit Compare

Figure 10.7 Two Hands of Cards

Source: TokenPhoto/iStock.com

This two-player game has the players draw 3 number cards from 10 possible numerals (see Figure 10.7). Each player creates the greatest possible two-digit number from the three cards and then compares their two-digit number with their partner's number. Students then write down both numbers on the playing sheet and use the symbols >, =, and < to compare the two numbers.

TASK PREPARATION

- Provide baggies with 10 number cards (0–9) for each student pair and a gameboard.

- This card game can be played with any number of pairs of students. It could be used as a center, small-group activity, or for the entire class.

LAUNCH

1. Say, "Let's play a game!"

2. Introduce the game to the class by selecting a student and playing a couple of rounds in front of the class on the document camera. Or, use large digit cards and post them so the whole class can see.

3. Say, "Each of us will draw 3 cards out of our baggie. Next, we decide which 2 cards we want to use to make the greatest number possible. Then, we write the symbol that shows how to compare our numbers." See Figure 10.8.

Mathematics Standard

- Understand that the two digits of a two-digit number represent amounts of tens and ones.

Mathematical Practice

- Reason abstractly and quantitatively.

Vocabulary

- greater than (>)
- less than (<)
- equal to (=)

Materials

- baggies with 10 number cards (0–9) for each pair of students
- Digit Compare student page gameboard

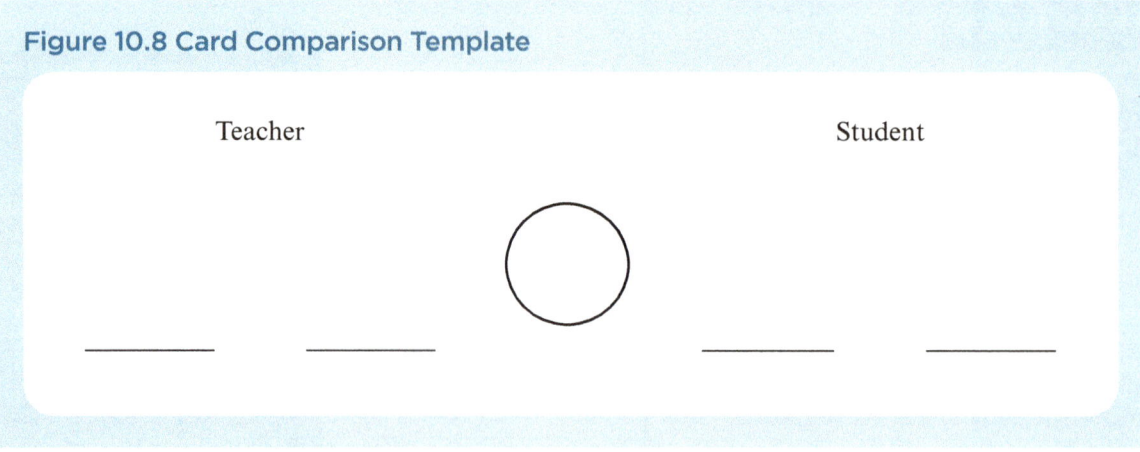

Figure 10.8 Card Comparison Template

Teacher   Student

4. Ask, "Who won that round?"

FACILITATE

1. **Observe** students as they consider the different combinations when making their two-digit number from the three cards. Ask, "How did you decide that was the greatest number? What do you know about tens and ones that helped you decide that was the greatest number you could make?"

> **! PRODUCTIVE STRUGGLE**
>
> Before asking questions, make sure to give students time to try out different combinations first. Some students may still be developing their understanding of counting by tens.

2. Using Show Me, ask students how many tens and how many ones are in their number.

3. **Interview.** Ask students to present their strategies in deciding which number is greater. Note all strategies suggested for sharing in the Close.

4. Ask, "Could the two numbers being created be equal? Explain why or why not."

Note: Consider using the Observation, Show Me, and Interview task tools to monitor student understanding (see Appendix B).

CLOSE: MAKE THE MATH VISIBLE

1. Have students share their strategies and record all the strategies they used when creating the greatest number.

2. Ask student pairs to share examples of one of their rounds of play.

3. **Hinge Question.** For this game, what strategy will always work when trying to make the greatest two-digit number? (using the largest number to represent the number of tens).

>  **STRENGTHS SPOTTING**
>
> Vary the ways that students share their strengths and use evidence of student thinking.

TASK 29: DIGIT COMPARE STUDENT PAGE GAMEBOARD

Task 29: Digit Compare Student Page

Digit Compare Gameboard

Instructions

1. Each player reach into the bag of number cards and select three cards.
2. Each player should now make the greatest two-digit number they can with two of their number cards and place the two cards in front of them.
3. Write down each two-digit number under each player's name on this pages and use (>, =, <) to compare the two numbers.
4. Check each other's two-digit numbers. Did you each make the greatest number possible? If not, what greater number could you have made?

   • Did you compare them correctly?
   • What strategies are you using?

Player 1 name: _____   Player 2 name: _____

Example: 35   <   49

_____   _____   _____

_____   _____   _____

_____   _____   _____

_____   _____   _____

POST-TASK NOTES: REFLECTION & NEXT STEPS

Mathematics Standard

- Compare two two-digit numbers based on meanings of the tens and ones digits, recording the results of comparisons with the symbols >, =, and <.

Mathematical Practices

- Reason abstractly and quantitatively.
- Construct viable arguments and critique the reasoning of others.

Vocabulary

- tens
- ones
- dimes
- pennies

Materials

- dimes
- pennies
- markers
- markerboards

Task 30
Coin Collections

Critique reasoning in a place value comparison task

TASK

Coin Collections

Figure 10.9 Coins for Comparison

Source: Penny by MisterVector/iStock.com; Dime by KavalenkavaVolha/iStock.com

If you want to have as much money as possible, would you rather have 2 dimes and 9 pennies, or 3 dimes and 1 penny (see Figure 10.9)? Shane says, "I would rather have 2 dimes and 9 pennies, because there are more coins." Elena disagrees with Shane and says, "No. I would rather have 3 dimes and 1 penny." Do you agree with Shane or Elena? Why?

TASK PREPARATION

- Students will complete the task individually and then in pairs.

- Students should be provided with either actual dimes and pennies or coin manipulatives.

- This task is designed to offer the class an opportunity to discuss different strategies they can use to compare two collections of coins.

ALTERNATE LEARNING ENVIRONMENT

If coins and physical manipulatives are not available, consider using a digital resource such as the money pieces virtual manipulative from The Math Learning Center: www.mathlearningcenter.org/resources/apps/money-pieces

LAUNCH

1. Gather the class together and present the following task:

    If you want to have as much money as possible, would you rather have 2 dimes and 9 pennies, or 3 dimes and 1 penny? Shane says, "I would rather have 2 dimes and 9 pennies, because there are more coins." Elena disagrees with Shane and says, "No. I would rather have 3 dimes and 1 penny."

2. Ask the class, "Do you agree with Shane or Elena? Why?"

3. Have students *Turn and Talk* to a partner, then have students share out their thoughts with the class.

4. Have the students vote to agree with Shane or Elena and record the votes with tally marks on the board.

FACILITATE

1. Have the students work individually, then pair them up to discuss their responses to the task. **Observe** the pair discussions.

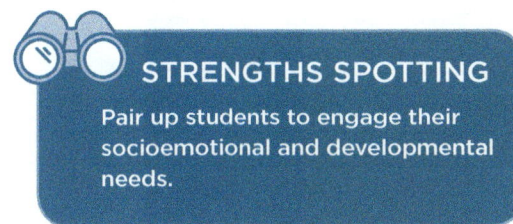

STRENGTHS SPOTTING

Pair up students to engage their socioemotional and developmental needs.

2. Ask students to present the strategies they used in deciding who has more money.

3. **Hinge Question.** What is a reason that Elena might disagree with Shane?

CLOSE: MAKE THE MATH VISIBLE

1. As students share their strategies to determine the total value of each group of coins, ask them to show you they actually determined the total amount.

2. Highlight the following ideas:

   » How many coins did Elena and Shane have each? What is the value of a dime? How does that compare to the value of a penny?

   » Does the number of coins that you have determine the value?

   » What is most important when determining the total value of a coin collection? The total number of coins or the value of each coin?

3. Ask the students, "What if one of you said that the larger number of dimes is the same as having a larger group of tens when comparing two two-digit numbers? Does that make sense? Why?"

POST-TASK NOTES: REFLECTION & NEXT STEPS

Task 31
Number Disagreement

Critique reasoning in a multidigit number comparison task

TASK

Number Disagreement

Figure 10.10 Mariana and Sean See the Same Number

Source: Children faces by Sudowoodo/iStock.com; Chalkboard by Sonya_illustration/iStock.com

Mariana and Sean are both looking at a number (see Figure 10.10). When asked what the number is, Mariana answers, "The number is twenty-three. There are two tens and three ones." Sean says, "I think the number is two and three." Do you agree with Mariana or Sean? How can you prove who is right?

TASK PREPARATION

- Students will complete the task in pairs. Consider how you will most effectively pair up your students.

LAUNCH

1. Gather the class together and provide the following task:

    Mariana and Sean are both looking at a number. When asked what the number is, Mariana answers, "The number is twenty-three. There are two tens and three ones." Sean says, "I think the number is two and three because I see a number 2 and a number 3."

2. Ask the class, "Do you agree with Mariana or Sean? Why or why not?"

3. Have students *Turn and Talk* to a partner, then have students share out their thoughts with the class.

4. Have the students vote to agree with Mariana or Sean and record the votes with tally marks on the board.

5. Have the students work individually, then pair them up to discuss their responses to the task. Ask, "What materials do you need to prove your idea?" **Observe** the pair discussions. Note how students are building representations of the number values.

6. **Interview.** Ask, "What does the '2' in 23 represent? How can you prove to Sean the value of the 2?"

7. Bring the class back together to share ideas. Using the students' work as a foundation, ask students to show how they built the 23 using counters or cubes. Ask students to present the strategies they used in deciding who is right.

8. **Hinge Question.** What is a reason that Mariana might disagree with Sean?

Note: Consider using the Interview tool to record student responses (see Appendix B).

FACILITATE

1. Arrange students in heterogeneous pairs and distribute Place Value Disagreement student pages. Have students work in pairs to work out which child is correct for each card.

2. **Observe/Interview.** As you circulate from pair to pair, take note of students' reasoning.

3. Ask:

   » Who do you think is correct?

   » How can you prove that they are correct using your counters or base ten blocks?

   » How does each student see the number?

Note: Consider using the Observation and Interview (individual or small group) tools (see Appendix B).

CLOSE: MAKE THE MATH VISIBLE

1. Bring the class back together. For each card, have the class vote for who they think is correct and record the results on the board.

2. Ask pairs to show how they proved which student was correct.

3. Highlight the following ideas:

   » Two-digit numbers are composed of tens and ones.

   » Changing the position of a digit (e.g., from ones to tens or tens to ones) changes its value.

4. Ask the students to respond to the following: "If I have the number 26 (circle the 2), how many does this represent?"

TASK 31: PLACE VALUE DISAGREEMENT STUDENT PAGES

 To download printable resources for this task, visit **resources.corwin.com/ ClassroomReadyMath/K-1**

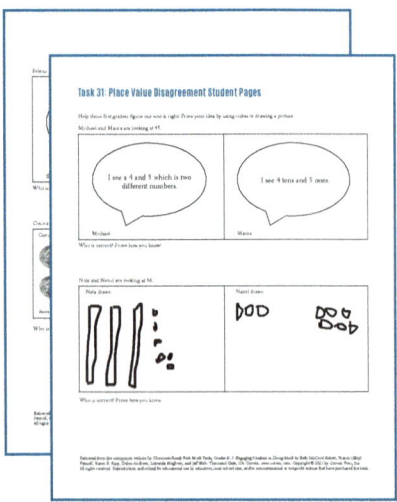

POST-TASK NOTES: REFLECTION & NEXT STEPS

CHAPTER

11

Number and Operations in Base Ten

Adding and Subtracting

TASK 32: GRADE 1: JUMPING AROUND ON A 120 CHART

Given a two-digit number, mentally find 10 more or 10 less than the number, without having to count; explain the reasoning used.

TASK 33: GRADE 1: TWO DIFFERENT STRATEGIES FROM DARLA AND JULIO

Add within 100, including adding a two-digit number and a one-digit number, and adding a two-digit number and a multiple of 10, using concrete models or drawings and strategies based on place value, properties of operations, and/or the relationship between addition and subtraction; relate the strategy to a written method and explain the reasoning used. Understand that in adding two-digit numbers, one adds tens and tens, ones and ones; and sometimes it is necessary to compose a ten.

TASK 34: GRADE 1: ROCKS FOR A ROCK COLLECTION

Add within 100, including adding a two-digit number and a one-digit number, and adding a two-digit number and a multiple of 10, using concrete models or drawings and strategies based on place value, properties of operations, and/or the relationship between addition and subtraction; relate the strategy to a written method and explain the reasoning used. Understand that in adding two-digit numbers, one adds tens and tens, ones and ones; and sometimes it is necessary to compose a ten.

Given a two-digit number, mentally find 10 more or 10 less than the number, without having to count; explain the reasoning used.

TASK 35: GRADE 1: TENS AWAY!

Subtract multiples of 10 in the range 10–90 from multiples of 10 in the range 10–90 (positive or zero differences), using concrete models or drawings and strategies based on place value, properties of operations, and/or the relationship between addition and subtraction; relate the strategy to a written method and explain the reasoning used.

Anticipating Student Thinking: The previous chapter's tasks, which all emphasized foundational concepts related to place value, provide the background necessary for this chapter's focus on adding and subtracting with two-digit numbers. Additionally, two of the chapter's tasks involve mental mathematics, as students determine 10 more or 10 less than a number. The instructional emphasis in each of the first-grade task lessons provided in this chapter is understanding the addition and subtraction process involving two-digit numbers. Each task will involve your students in using particular representational tools (e.g., base ten blocks, 120 charts) as they add or subtract. Use of Hinge Questions, Interviews, and Show Me, as classroom-based formative assessment techniques, will provide feedback to your students and will also serve as a guide for your next steps instructionally.

THINK ABOUT IT

As you plan for and implement the chapter's task lessons, consider how you may communicate what your students are learning to their parents and family members. It's important for them to know that your class is working hard to understand how addition and subtraction work conceptually. Procedures come later.

Mathematics Standard

- Given a two-digit number, mentally find 10 more or 10 less than the number, without having to count; explain the reasoning used.

Mathematical Practices

- Use appropriate tools strategically.
- Look for and make use of structure.

Vocabulary

- addition
- subtraction
- 120 chart

Materials

- 120 charts (large one for class viewing, and individual 120 charts student page)
- base ten blocks

Task 32
Jumping Around on a 120 Chart

Use a 120 chart to add and subtract tens

TASK

Jumping Around on a 120 Chart

Figure 11.1 Abu's and Annie's Jumps

111	112	113	114	115	116	117	118	119	120
102	102	103	104	105	106	107	108	109	110
91	92	93	94	95	96	97	98	99	100
81	82	83	84	85	86	87	88	89	90
71	72	73	74	75	76	77	78	79	80
61	62	63	64	65	66	67	68	69	70
51	52	53	54	55	56	57	58	59	60
41	42	43	44	45	46	47	48	49	50
31	32	33	34	35	36	37	38	39	40
21	22	23	24	25	26	27	28	29	30
11	12	13	14	15	16	17	18	19	20
1	2	3	4	5	6	7	8	9	10

Ms. Cook asked her class to use a 120 chart to find 40 less than 68. Abu says: "I know how to do that! I start at the 68 on the 120 chart and I jumped 4 to the left." Annie says: "I solved it a different way. I also started at the 68, and I jumped 4 down." Figure 11.1 shows the jumps described by Abu and Annie. Who do you think is right: Abu or Annie? Why?

TASK PREPARATION

- This task engages students in using a 120 chart. The 120 chart presented is sometimes called "bottom up" because the bottom row starts with 1–10. The structure of the chart is beneficial to children because of the connection between adding and subtracting based on place value.

Adding or subtracting 10 (vertical movement) changes the value of the tens place digit, while adding or subtracting 1 (horizontal movement) changes the value of the ones place digit. The 120 chart also allows children to see each of the tens as 10 ones or 1 ten, and that 100 is 10 tens. With a bottom-up chart, children also see that when a number gets larger, it moves up the chart.

- Make sure that each student has their own 120 chart.

LAUNCH

1. Say, "Today we are going to learn about a new math tool!" (Show the class a large 120 chart.) "What do you notice about the numbers on this chart? What do you wonder about the chart? What patterns do you see?"

2. Say, "Here is our first problem: Ms. Cook asked the class to use the 120 chart to figure out what is 10 more than 15. Brett started at 15 and jumped once to the right. Latoya started at 15 and jumped one up. Figure 11.s shows their respective jumps in the 120 chart. Who solved it correctly, Brett or Latoya?"

STRENGTHS SPOTTING

Recognize students' contributions as strengths. Children are naturally drawn to patterns and will be excited to share the patterns they see. Encourage, recognize, and celebrate students' curiosity as they ask questions (wonders). Record questions to revisit in later lessons.

Figure 11.2 Brett's and Latoya's Jumps

41	42	43	44	45	46	47	48	49	50
31	32	33	34	35	36	37	38	39	40
21	22	23	24	25	26	27	28	29	30
11	12	13	14	15	16	17	18	19	20
1	2	3	4	5	6	7	8	9	10

41	42	43	44	45	46	47	48	49	50
31	32	33	34	35	36	37	38	39	40
21	22	23	24	25	26	27	28	29	30
11	12	13	14	15	16	17	18	19	20
1	2	3	4	5	6	7	8	9	10

3. Ask students to *Turn and Learn* from a partner: "Show the number 15 on your 120 chart to your partner. Discuss how you would solve the problem: what is 10 more than 15?" Then tell them to decide how they know who solved it correctly, Brett or Latoya.

4. Ask several pairs to share their solution strategies and how they decided who solved it correctly. Ask what jumping to the right means. Ask what jumping up one box means when you are adding 10. Have them demonstrate why this works.

5. Share the task with the class:

> Ms. Cook asked her class to use a 120 chart to find 40 less than 68. Abu says: "I know how to do that! I start at the 68 on the 120 chart and I jumped 4 to the left." Annie says: "I solved it a different way. I also started at the 68, and I jumped 4 down."

> Who do you think is right: Abu or Annie? Why?

FACILITATE

1. As students begin to work individually, use **Observation** to monitor their progress. Consider the following "look fors":

   » Are students able to accurately solve problems using the 120 chart?

   » Do they correctly interpret the meaning of jumping to the right?

   » Can they explain why they are jumping up for adding each 10?

PRODUCTIVE STRUGGLE

As you observe, resist the urge to intervene and make corrections. Use the opportunity to attend to student thinking with genuine curiosity and interest. Pay close attention to how students work through misconceptions that pop up. Look for strengths that can be leveraged to help the whole class move forward, particularly those students who see patterns when jumping up or to the right.

2. Your **Observations** may prompt you to consider conducting individual student **Interviews**. Ask:

   » Why do two jumps up mean adding 20 more? Show me why this works. What do two jumps down mean?

   » What does one jump to the right mean? What happens to the number when you make one jump to the right? To the left?

3. Also consider using **Show Me**, with individual students, asking, "Show me what you would do to add 30 more? What would you do to subtract 40 less?"

4. After students complete the first task, ask them to use their 120 chart to find

   a. 30 more than 51,

   b. 40 less than 53, and

   c. 50 less than 72, using at least two different strategies.

Note: Consider using the Observation, Interview (individual and small group), and Show Me tools to monitor and record student comments (see Appendix B).

CLOSE: MAKE THE MATH VISIBLE

1. Bring the class back together to share responses to the **Interview** questions.

2. Focus the discussion on the meaning of jumping up one on the chart to add 10 and on a student solution of finding 10 more than a number without having to count. Also discuss how jumping to the right 10 times relates to jumping up. Or jumping to the left 10 times relates to jumping down one.

3. **Hinge Question.** Ask the class to show 30 more than 71 on the 120 chart.

TASK 32: JUMPING AROUND ON A 120 CHART STUDENT PAGES

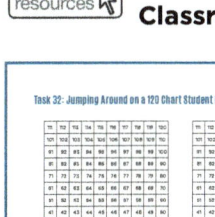 To download printable resources for this task, visit **resources.corwin.com/ ClassroomReadyMath/K-1**

Task 32: Jumping Around on a 120 Chart Student Page

POST-TASK NOTES: REFLECTION & NEXT STEPS

Mathematics Standard

- Add within 100, including adding a two-digit number and a one-digit number, and adding a two-digit number and a multiple of 10, using concrete models or drawings and strategies based on place value, properties of operations, and/or the relationship between addition and subtraction; relate the strategy to a written method and explain the reasoning used. Understand that in adding two-digit numbers, one adds tens and tens, ones and ones; and sometimes it is necessary to compose a ten.

Mathematical Practices

- Look for and make use of structure.
- Construct viable arguments and critique the reasoning of others.

Vocabulary

- compensation

Materials

- base ten blocks
- bags of at least 50 counters
- Two Different Strategies student page

Task 33

Two Different Strategies From Darla and Julio

Add two two-digit numbers using tens and ones strategies

TASK

Two Different Strategies From Darla and Julio

Figure 11.3 Adding 18 and 25

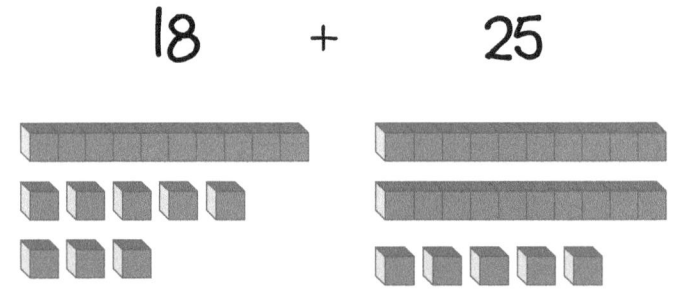

Darla and Julio were both adding 18 and 25 (see Figure 11.3). Darla said: "I'm going to count the tens first from the 18 and the 25. 10, 20, 30. And then I'm going to count the ones: 8, 9, 10, 11, 12, 13. So 30 plus 13 equals 43."

Julio said: "That's a really great strategy, Darla! Want to hear how I did it? I took 2 from the 25 and gave it to the 18. The 18 is now 20, and the 25 is now 23. So 20 plus 23 equals 43. I got the same answer as you!"

Can you explain how Darla and Julio solved their problems? Will their strategies always work? How do you know?

TASK PREPARATION

- Organize students into heterogeneous groups of two or three.

- Prepare base ten blocks for those students who may benefit from using concrete materials.

- Provide bags of at least 50 or so counters for each student work group.

ALTERNATE LEARNING ENVIRONMENT

If physical manipulatives are not available, consider using a digital resource such as the number pieces app from The Math Learning Center: apps.mathlearningcenter.org/number-pieces/

LAUNCH

1. Start the students with the problem. Say, "Lamar had 18 counters. He gathered 25 more. How many counters does Lamar have now? Turn and talk with your neighbor. Using your group's bag of counters, find and share different ways that you might solve this problem."

2. Record student strategies on the board. Highlight strategies that are mathematically different, especially those that use compensation in different ways. Julio's strategy is an example of a compensation strategy, where he takes 2 from the 25 to make the 18 a 20. Then, he adds the 20 to the 23 to get 43. There are many different ways that students use compensation to create "friendly" numbers. Another common example would be a student adding 20 to 25 instead of 18 + 25. The student would add 20 + 25 = 45 and then subtract the 2 at the end, so 45 – 2 = 43.

FACILITATE

1. Read and act out the task for the students:

   "Darla and Julio from another class were solving the same math problem. Darla said, 'I'm going to count the tens first from the 18 and the 25. 10, 20, 30. And then I'm going to count the ones: 8, 9, 10, 11, 12, 13. So 30 plus 13 equals 43.'" (See Figure 11.4.)

Figure 11.4 Darla's Strategy

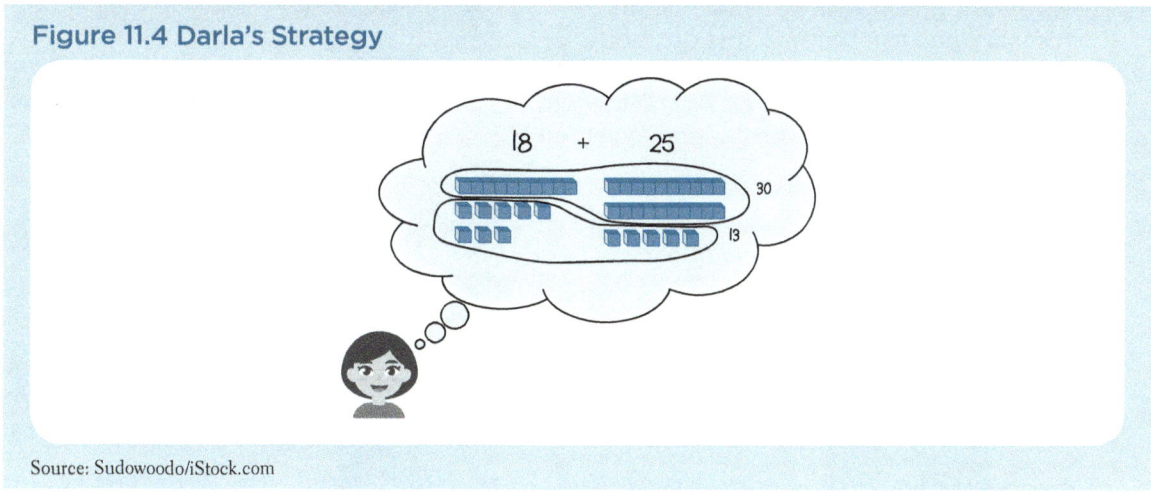

Source: Sudowoodo/iStock.com

   "Julio said, 'That's a really great strategy, Darla! Want to hear how I did it? I took 2 from the 25 and gave it to the 18. The 18 is now equal to 20, and the 25 is now 23. So 20 plus 23 equals 43. I got the same answer as you!'" (See Figure 11.5.)

Figure 11.5 Julio's Strategy

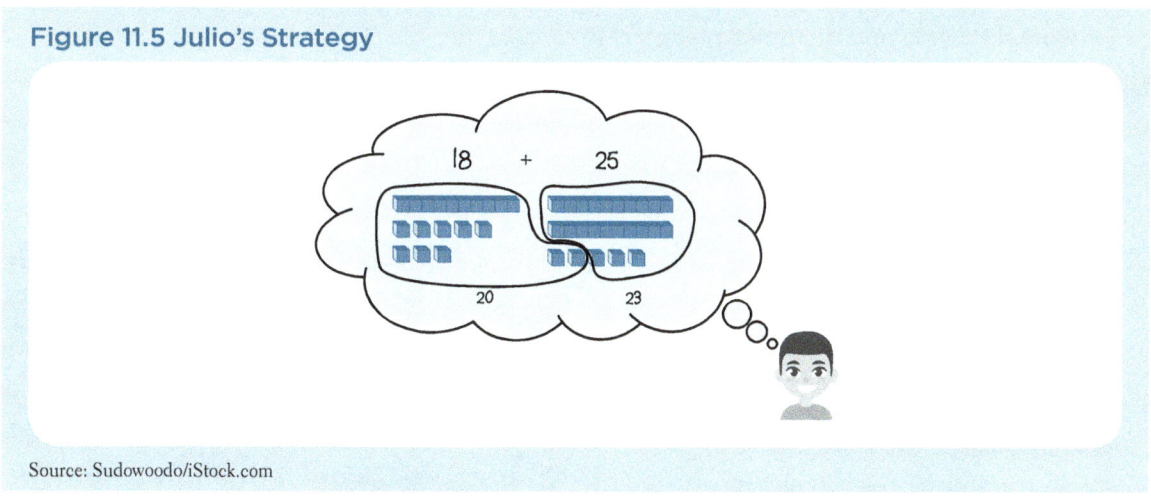

Source: Sudowoodo/iStock.com

2. As you verbally explain their solution strategies, draw them on the board (Figure 11.6).

Figure 11.6 Illustrating Darla's and Julio's Strategies

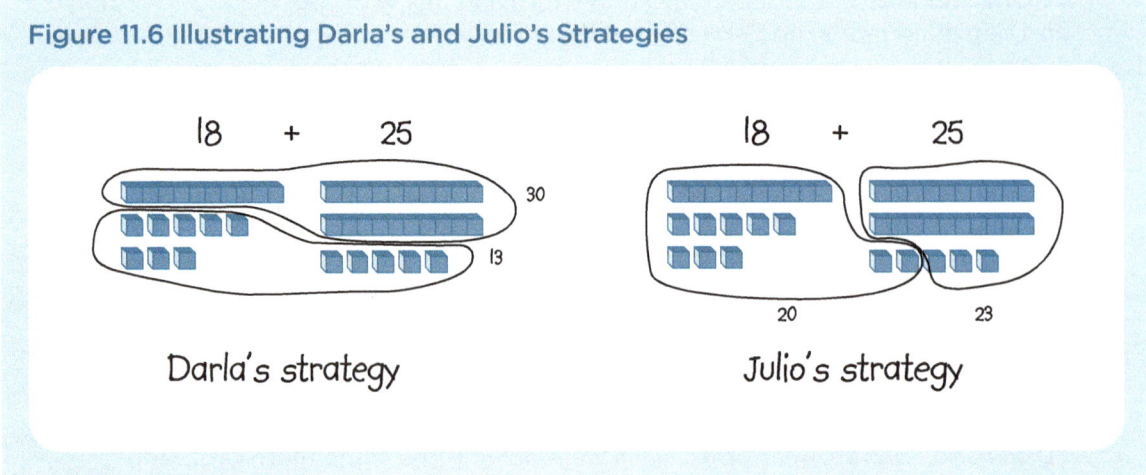

Darla's strategy

Julio's strategy

3. Organize students into pairs or small groups.

4. Provide each group with base ten blocks if necessary and ask them to test out both methods (Darla's way and Julio's way) as well as their own solution strategies. Ask, "How did Darla and Julio solve their problems? Will their strategies always work?"

5. As you visit each group, use **Show Me** and targeted small-group **Interview** questions. Ask:

   » Show me how you would use Darla's strategy to solve 18 + 25.

   » Show me how you would use Julio's strategy to solve 18 + 25.

   » Whose strategy do you prefer? Or do you prefer your own? Show me your strategy, if it's different from Darla's and Julio's.

   » How are their strategies similar to or different from your strategies?

   » How did Darla and Julio use place value or base ten blocks to solve the problem?

Note: Consider using the Interview (small group) and Show Me tools for documenting progress (see Appendix B).

> ### ! PRODUCTIVE STRUGGLE
>
> Students may share that they do not understand Darla's and Julio's strategies and that they are comfortable with their own strategies. Celebrate the strength in this thinking—it is a good thing to have your own strategy that makes sense to you. Build on the idea that understanding other people's strategies is important because there are many different ways to solve problems as well as the idea that their strategy might be similar to a different strategy or even one you like better.

CLOSE: MAKE THE MATH VISIBLE

1. After the students have had time to try both Darla's and Julio's strategies and perhaps generate their own, bring the class back together. Using comments gathered during the Show Me/interviews, ask students to share evidence of their own thinking.

> ### STRENGTHS SPOTTING
>
> Encouraging students to represent their solutions in different ways provides teachers an opportunity to notice and leverage representations that are strengths for each student.

2. Using base ten blocks, demonstrate the following strategies to the students:

» Darla added all of the tens together first and then added all of the ones.

» Julio broke apart 25 to make the 18 into a 20, which was easier for him to add.

» Another common strategy is where a student starts with one number and adds the other on in pieces. In this case, a student might start with the 18 and add the two tens from the 25 and then the last 5: 18, 28, 38, 39, 40, 41, 42, 43. Or start with the 25 and add the 8 from the 18, and then the 10: 25, 26, 27, 28, 29, 30, 31, 32, 33, 43.

» Emphasize to the students that both of these strategies used place value understanding, working with tens and ones in different ways. Ask students if they can explain why both strategies work.

TASK 33: TWO DIFFERENT STRATEGIES STUDENT PAGE

 To download printable resources for this task, visit **resources.corwin.com/ ClassroomReadyMath/K-1**

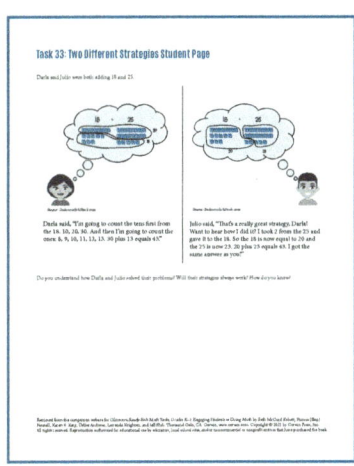

POST-TASK NOTES: REFLECTION & NEXT STEPS

Mathematics Standards

- Add within 100, including adding a two-digit number and a one-digit number, and adding a two-digit number and a multiple of 10, using concrete models or drawings and strategies based on place value, properties of operations, and/or the relationship between addition and subtraction; relate the strategy to a written method and explain the reasoning used. Understand that in adding two-digit numbers, one adds tens and tens, ones and ones; and sometimes it is necessary to compose a ten.

- Given a two-digit number, mentally find 10 more or 10 less than the number, without having to count; explain the reasoning used.

Mathematical Practices

- Make sense of problems and persevere in solving them.
- Look for and make use of structure.

Vocabulary

- estimate
- mentally adding 10
- sum

Task 34
Rocks for a Rock Collection

Understand that there are multiple ways to add two-digit numbers

TASK

Rocks for a Rock Collection

Figure 11.7 Dorothy's Bag of Rocks

Source: Paper bag by aleksei-veprev/iStock.com; Stones by pixabay.com and Pobytov/iStock.com.

Dorothy had an entire bag full of rocks for her rock collection (Figure 11.7). As she was walking across the playground, she tripped and the bag opened up. Her rocks landed everywhere! When she got home again, she counted the rocks that were still in her bag. She had 62. When her bag is completely full, it can hold 100 rocks. How many rocks does she need to refill the bag?

TASK PREPARATION

- Organize students into heterogeneous pairs.

LAUNCH

1. Show the class a bag filled with 100 rocks or a picture of a bag of rocks.

2. Ask the class to estimate how many rocks there are in the bag.

3. Record the estimates on sticky notes and post them on the board from least to greatest. Reveal to the class that there are 100 rocks in the bag when it is completely full.

FACILITATE

1. Present the task to the class, and have students *Turn and Talk* with a partner. Ask, "How many rocks could Dorothy add to her bag so that it is full with 100 rocks? How do you know?"

2. Distribute a task recording sheet, little ten frames, and a 120 chart to each pair.

3. Direct the student pairs to work together as follows:

   » Start with 98 rocks and find out how many rocks they need to add to Dorothy's bag to fill it (to 100).

   » Choose a different number of rocks to start with (e.g., 45) and repeat the process.

ACCESS AND EQUITY

Provide enough time for students to work through at least two different starting numbers as certain numbers lend themselves better to different strategies. Consider letting students choose numbers as choice supports students to feel more empowered to solve the problem.

4. As students work, circulate and **Interview** each student pair, asking:

   » What is your strategy for choosing the number to add to your starting number?

   » How is your addition strategy the same as or different from your partner's strategy?

   » What tool (120 chart, base ten blocks, little ten frames) are you using to help you solve?

   » How did _____'s strategy help you?

   » What should you do if your answer (sum) is greater than 100?

STRENGTHS SPOTTING

Incorporate questions like these that cultivate strength building in students. Help them see themselves as mathematicians and encourage them to share their own mathematical strengths as well as the strengths they see in others!

Note: Consider using the Interview (small group) tool to collect student responses (see Appendix B).

Materials

- sticky notes
- counters
- base ten blocks
- 120 Chart student page
- little ten frames, one for each student
- Rock Collection student page
- Bags of Rocks student page

CLOSE: MAKE THE MATH VISIBLE

1. Once each pair has finished their recording sheet, facilitate a Notice and Wonder Gallery Walk. Make sure that students discuss the strategies that were used as well as the numbers that were chosen by the other pairs.

2. Select students to share their strategies based on the interviews conducted during the Facilitate phase of the task lesson. Highlight strategies that used place value in different ways (adding the tens and then adding the ones, composing a 10 from 10 ones, decomposing a 10 to make the other number easier to add).

3. Encourage the students to share how their strategies changed depending on the starting number.

STRENGTHS SPOTTING

When students articulate how their thinking has changed or grown, focus on the strengths they share by naming the strength. For example, perhaps they tried multiple solution pathways. When students understand how they used their strengths to revise learning or build new learning, they are more likely to exhibit this behavior again.

TASK 34: ROCKS FOR A ROCK COLLECTION STUDENT PAGES

online resources | To download printable resources for this task, visit **resources.corwin.com/ ClassroomReadyMath/K-1**

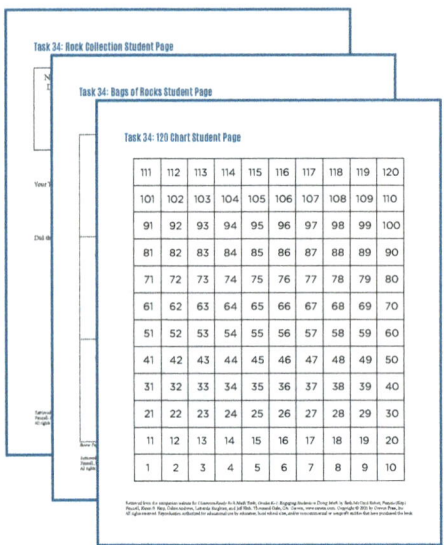

POST-TASK NOTES: REFLECTION & NEXT STEPS

Task 35
Tens Away!

Demonstrating place value understandings, subtracting multiples of 10

TASK

Tens Away!

Tim, a first grader in another class, made the following conjecture:

"If I use a 120 chart, start at 80 and count backwards by 10, I will never land on a number that has a 9 in the ones place." (See Figure 11.8.)

Figure 11.8 Testing Tim's Conjecture

87	88	89	90
77	78	79	80
67	68	69	70
57	58	59	60
47	48	49	50
37	38	39	40

Is Tim's conjecture true or false? Could Tim end up on a number that has a 9 in the ones place if he counts backwards by 10? Why or why not?

TASK PREPARATION

- This task starts with a game and then moves on to the task.

- The task is meant to be solved in pairs, so consider how you will pair up your students.

- Think about not only which students are comfortable with the mathematics of subtracting multiples of 10, but also which students have strengths in explaining their reasoning and justifying their thinking.

- Each pair should have access to a 120 chart, Decade Spinner, and a Tens Away! recording sheet.

Mathematics Standard

- Subtract multiples of 10 in the range 10–90 from multiples of 10 in the range 10–90 (positive or zero differences), using concrete models or drawings and strategies based on place value, properties of operations, and/or the relationship between addition and subtraction; relate the strategy to a written method and explain the reasoning used.

Mathematical Practices

- Construct viable arguments and critique the reasoning of others.

- Look for and express regularity in repeated reasoning.

Vocabulary

- conjecture
- counting backwards
- subtraction
- zero
- ones place

Materials

- 120 Chart student page
- Decade Spinner student page
- base ten blocks
- Tens Away! student page

LAUNCH

1. Gather the whole class and note: "Class, today you are going to play a game with a partner called Tens Away! The first person is going to find 90 on the 120 chart. The second person is going to spin the Decade Spinner and subtract that amount from 90. The first person is then going to write an equation to record what happened."

2. Say, "Let's go through the game once to make sure we all understand how to play the game."

3. Say, "I'm going to start at the number 90" (point to 90 on the 120 chart).

4. Pick a student to spin the spinner. If the spinner lands on 30, say, "Because the spinner landed on 30, I am going to remove or subtract 30 from my original number of 90."

5. Have the students move from the whole group to working pairs.

6. Ask students to *Turn and Talk* with a neighbor to discuss different ways to solve the problem.

7. Ask several students to share their strategies, with specific requests for students to show how they solved the problem using the 120 chart, base ten blocks, or counting back by tens.

8. Write the equation 90 – 30 = 60 on the recording sheet. Ask the student pairs to switch roles and play two more rounds of the game.

FACILITATE

1. As the students begin to play Tens Away! in pairs, observe, specifically looking for the following:

   » What math tools are the students using?

   » Are they counting by tens to solve the problem, or do they still need to count by ones?

   » Are students using models or mentally counting backwards?

> **PRODUCTIVE STRUGGLE**
>
> Some students will feel more comfortable counting by ones or may need the base ten blocks. This game provides an opportunity for them to use the strategies they are comfortable with and see strategies that use tens and are more abstract.

2. As you **Observe** and monitor task progress, **Interview** the pairs. Consider asking:

   » Is it faster to count backwards from 80 by tens or by ones? Why?

   » What happens to the tens digit when you count backwards from 80?

   » What happens to the ones digit?

   » What do you know about counting tens that helps you to solve this problem?

3. Present the task to the student pairs:

   » Tim, a first grader in another class, made the following conjecture:

   "If I use a 120 chart, start at 80 and count backwards by 10, I will never land on a number that has an 9 in the ones place."

   » Is Tim's conjecture true or false? Could Tim end up on a number that has an 9 in the ones place if he counts backwards by 10? Why or why not?

4. Once the student pairs respond to the conjecture, ask them to decide if it was true or false and if Tim could end up with a number that has a 9 in the ones place, making sure they can justify why or why not.

Note: Consider using the Observation and Interview (small group) tools for recording student responses (see Appendix B).

STRENGTHS SPOTTING

Some students will rely on tools such as the 120 chart or base ten blocks to solve the problem. Acknowledge and value their successful solution strategies, while encouraging them toward subtraction and counting back by tens.

CLOSE: MAKE THE MATH VISIBLE

1. Once each pair has played at least three rounds of the Tens Away! game and responded to the accuracy of Tim's conjecture, bring the class back together. Have the class respond to the following questions:

   » When you played the game and completed the counting related to Tim's conjecture, what happened to the ones digit when you subtracted tens?

   » What happened to the tens digit when you subtracted tens?

   » Did you notice any patterns? Show me an example of a pattern you saw.

2. Students should see multiple examples subtracting tens from 80, 90, and 100. In these examples, students should notice that the ones digit does not change when subtracting by tens.

3. **Exit Task.** Tim says that if he starts at 40 and adds tens, he will never end up with a number different from 0 in the ones place. Is Tim's statement true or false? How do you know?

TASK 35: TENS AWAY! STUDENT PAGES

 To download printable resources for this task, visit **resources.corwin.com/ClassroomReadyMath/K-1**

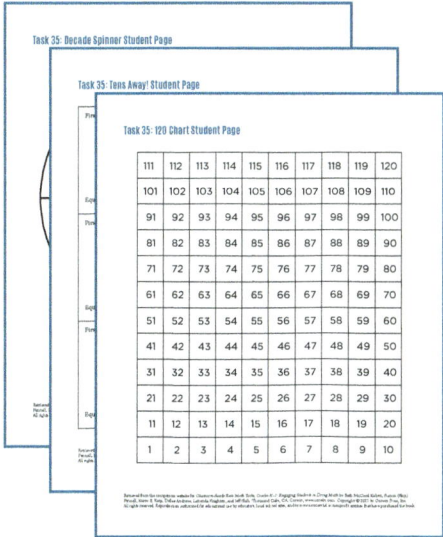

POST-TASK NOTES: REFLECTION & NEXT STEPS

Measurement

Comparing, Ordering, Sorting, and
Informal Measurement

TASK 36: KINDERGARTEN: WHICH IS HEAVIER?

Directly compare two objects with a measurable attribute in common, to see which object has "more of"/"less of" the attribute, and describe the difference. *For example, directly compare the heights of two children and describe one child as taller/shorter.*

TASK 37: GRADE 1: MAKE THE BEAR FAMILY

Order three objects by length; compare the lengths of two objects indirectly by using a third object.

Express the length of an object as a whole number of length units, by laying multiple copies of a shorter object (the length unit) end to end; understand that the length measurement of an object is the number of same-size length units that span it with no gaps or overlaps. *Limit to contexts where the object being measured is spanned by a whole number of length units with no gaps or overlaps.*

TASK 38: KINDERGARTEN: COMPARING CUPS

Directly compare two objects with a measurable attribute in common, to see which object has "more of"/"less of" the attribute, and describe the difference. *For example, directly compare the heights of two children and describe one child as taller/shorter.*

TASK 39: KINDERGARTEN: GUESS MY SORTING RULE

Classify objects into given categories based on their attributes; count the numbers of objects in each category and sort the categories by count.

TASK 40: GRADE 1: WHICH WORM IS WHICH?

Order three objects by length; compare the lengths of two objects indirectly by using a third object.

TASK 41: GRADE 1: MEASURE AND COMPARE

Express the length of an object as a whole number of length units, by laying multiple copies of a shorter object (the length unit) end to end; understand that the length measurement of an object is the number of same-size length units that span it with no gaps or overlaps. *Limit to contexts where the object being measured is spanned by a whole number of length units with no gaps or overlaps.*

Anticipating Student Thinking: The tasks provided in this chapter will engage your students in premeasurement and informal measurement activities. The chapter's tasks include directly comparing the size, weight, and attributes of objects at the kindergarten level (Tasks 36, 38, and 39), and ordering and comparing lengths and measuring lengths using non-standard units at the first-grade level (Tasks 37, 40, and 41). Measuring is truly doing mathematics. Adults estimate and measure throughout their lives. The tasks in this unit provide measurement process experiences that are important prerequisites to estimating and measuring using standard units of measure (beginning in Grade 2). As you observe your students engaging in the chapter tasks, notice how they begin to "see" lengths and naturally compare and order size and distances.

THINK ABOUT IT

As you prepare for the implementation of the tasks provided in this chapter, allot time for locating, disseminating, and retrieving the materials for the tasks.

Task 36

Which Is Heavier?

Compare measurable attributes

TASK

Which Is Heavier?

Sheena and Mark were helping Coach Nelson put some balls away after gym class. They had to put away ping pong balls, golf balls, and Wiffle balls (Figure 12.1).

- Coach asked them to put the heaviest balls on the bottom of the basket and the lighter balls on top. Mark said that because the Wiffle balls are larger than the ping pong balls and golf balls, they should go in the bottom of the basket. Do you agree or disagree with Mark? Explain your thinking.

- Mark and Sheena compared the weights of the balls in order to complete the task from Coach Nelson. If I had 10 Wiffle balls and 10 golf balls, which group of balls would be heavier?

Figure 12.1 Three Balls

| ping pong ball | golf ball | Wiffle ball |

Source: Ping pong ball and golf ball by Grafner/iStock.com; Wiffle ball by Glen Edwards/iStock.com

TASK PREPARATION

- In kindergarten, students begin to explore and describe measurable attributes of objects, which include weight/ mass. Common descriptive words used by students are tall, short, heavy, or light. Eventually, students transition to comparing two objects based on a measurable attribute. It is important that students have prior experiences with describing and exploring weight as a measurable attribute before engaging in this comparison task. In addition to comparing the weight/mass of objects, students are expected to use a simple balance scale to compare the mass of objects in this task. Before engaging in this task, provide opportunities for students to explore using a balance scale.

Mathematics Standard

- Directly compare two objects with a measurable attribute in common, to see which object has "more of"/"less of" the attribute, and describe the difference. *For example, directly compare the heights of two children and describe one child as taller/ shorter.*

Mathematical Practices

- Construct viable arguments and critique the reasoning of others.

- Use appropriate tools strategically.

Vocabulary

- heavy
- heavier
- heaviest
- light
- lighter
- lightest
- weight
- balance scale

Materials

- ping pong balls, golf balls, and Wiffle balls (4–5 of each if possible)
- Which Is Heavier? student page
- balance scale, one per group of students
- crayons, 10 per group of students
- craft sticks, 10 per group of students
- pennies, 10 per group of students
- counters, 10 per group of students

- Place students in groups of three for this task. Collect the materials for the measurement task and organize them for easy distribution to students.

LAUNCH

1. Distribute a golf ball, ping pong ball, and Wiffle ball to each table group. Tell students: "Examine the balls at your table. What do you notice? What do you wonder? Discuss your ideas with your tablemates." **Observe** the student groups as they discuss. Record student ideas. Ask, "How would you describe these balls? What are some ways we could figure out which one is heavier? Lighter?" Record student ideas.

Note: Use the Observation tool for recording student ideas (see Appendix B).

2. Share the beginning of the task with students. Say,

   "Sheena and Mark were helping Coach Nelson put some balls away after gym class. They had to put away ping pong balls, golf balls, and Wiffle balls. Coach asked that they put the heaviest balls on the bottom of the basket and the lightest balls on top."

3. Ask, "What do Sheena and Mark need to know to complete this task? Why would Coach want to put the heavier balls on the bottom? Turn and talk to a partner." (Students should identify weight as a measurable attribute. Students may also use the words *heavy* and *light* in their descriptions.)

4. Distribute a balance scale to each group and continue with the task. Say,

   "Mark said that because the Wiffle balls are larger than the ping pong and golf balls, they should go on the bottom. Do you agree or disagree with Mark? Turn and talk with a partner to explain why."

   Ask, "Can you show me how we could use the balls to help answer this question?" Allow time for students to discuss either with a partner or as table groups. Allow students to share their responses with the whole group. Students should be able to demonstrate (during the **Show Me**) and articulate that even though the Wiffle ball is larger in size, it is lighter than the golf ball so Mark is incorrect.

ACCESS AND EQUITY

Students who have never used a balance scale will want to experiment with the tool. Consider providing students with a brief exploration period where they choose objects to weigh on the balance scale to reacquaint themselves with the tool before starting the comparison task. The exploration time will also allow time for a quick review if needed.

NOTE: Consider using the Show Me tool for recording student responses (see Appendix B).

FACILITATE

1. Continue with the task. Say, "Mark and Sheena compared the weights of the balls in order to complete the task from Coach Nelson. If I had 10 Wiffle balls and 10 golf balls, which group of balls would be heavier?" Allow time for student responses. "You're correct, the golf ball is heavier than a Wiffle ball, so 10 golf balls are heavier than 10 Wiffle balls. Now, you're going to compare the weights of some classroom objects to determine which object is heavier than the other."

PRODUCTIVE STRUGGLE

Students may have trouble comparing Wiffle balls to golf balls without using the actual objects. Allow students to compare two of each type of ball, then three of each type, and so on up to five of each type before answering the question.

2. Divide students into groups of three; distribute the materials (balance scale, crayons, craft sticks, pennies, counters, and the Which Is Heavier? recording sheets); and go over the directions. **Observe** as students work to place objects on the balance scale. Make note of whether or not students use the balance scale correctly. Are students able to compare the weights of the objects and use measurement language in their comparisons? Can students articulate which objects are heavier?

3. Facilitate a *One Stay, Two Stray* to allow student groups to compare and discuss their findings. Encourage students to explain how they know an object is heavier. **Observe** as students share and ask questions as needed. Sample questions could include the following: "What did your group discover? Were there any surprises? How do you know which objects are heavier than the others?" Encourage students to check with the visiting group to see if they have the same answers.

Note: Consider using the Observation and Show Me tools for recording responses discussed above (see Appendix B).

CLOSE: MAKE THE MATH VISIBLE

1. Bring students together for a whole-group discussion. Allow students to share the results of their measurement comparisons. To facilitate the debriefing discussion, have students use the words *heavier* and *lighter* to describe the objects. Ask students to explain how they know one object is heavier than another object. ("What happened to the side of the balance scale that held the heavier objects? What happened to the side of the balance scale that held the lighter objects?")

2. Show students three new sports balls, such as a softball, a soccer ball, and a basketball, and ask, "Can we tell which of these balls is the heaviest based on the size of the ball?"

3. **Hinge Question.** Which ball is heaviest or lightest, and how do you know?

TASK 36: WHICH IS HEAVIER STUDENT PAGE

 To download printable resources for this task, visit **resources.corwin.com/ ClassroomReadyMath/K-1**

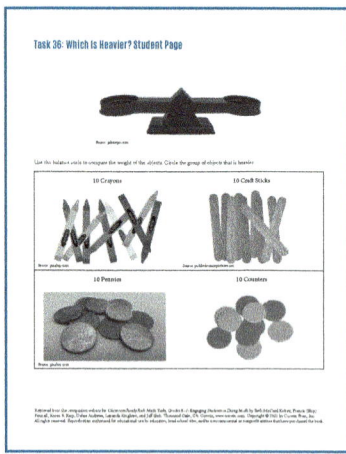

POST-TASK NOTES: REFLECTION & NEXT STEPS

Task 37

Make the Bear Family

Measure length indirectly and by iterating units

TASK

Make the Bear Family

Baby Bear is twice as tall as Goldilocks. Mama Bear is four times as tall as Goldilocks. Papa Bear is six times as tall as Goldilocks. Use the Goldilocks picture to find the heights of Baby, Mama, and Papa Bear. Then make each member of the Bear family. How tall is each member of the Bear family? Be prepared to show how you know.

Use the Goldilocks unit or the Baby Bear unit to make a tree for the forest where the bears live. Your tree has to be at least two Goldilocks units tall. How many Goldilocks units tall is your tree? How many Baby Bear units tall is your tree? Be prepared to show how you know how tall your tree is.

TASK PREPARATION

- This task requires students to use their prior experience with measuring length and should be introduced after students have had multiple opportunities to measure and compare lengths of a variety of objects. In this task, students will work in groups of three. Students will need access to a variety of materials to complete this task. To make distribution of materials more efficient, it may be helpful to assemble baggies of materials for each group ahead of time.

LAUNCH

1. Divide students into groups of three and distribute Cuisenaire rods (Figure 12.2). Ask, "What do we need to think about when measuring the length of an object?" Allow time for students to respond. Responses should include "line up units end to end without any gaps or overlaps." Say, "We can think of these as rules for measuring length. We're going to measure objects using the red rod as the unit."

Mathematics Standards

- Order three objects by length; compare the lengths of two objects indirectly by using a third object.

- Express the length of an object as a whole number of length units, by laying multiple copies of a shorter object (the length unit) end to end; understand that the length measurement of an object is the number of same-size length units that span it with no gaps or overlaps. *Limit to contexts where the object being measured is spanned by a whole number of length units with no gaps or overlaps.*

Mathematical Practices

- Make sense of problems and persevere in solving them.

- Attend to precision.

Vocabulary

- measure
- order
- height
- more
- shorter than
- gap
- overlap
- taller than

Materials

- baggie of red, purple, dark green, brown, and orange Cuisenaire rods only, one per group of three students
- Make the Bear Family student page, one Goldilocks page and one Make a Tree page per group of three students
- Make the Bear Family Goldilocks Cutouts student page, multiple copies for each group
- construction paper
- glue sticks
- scissors
- pencils and/or markers
- chart paper or poster paper, one per small group

Figure 12.2 Measuring Using Cuisenaire Rods

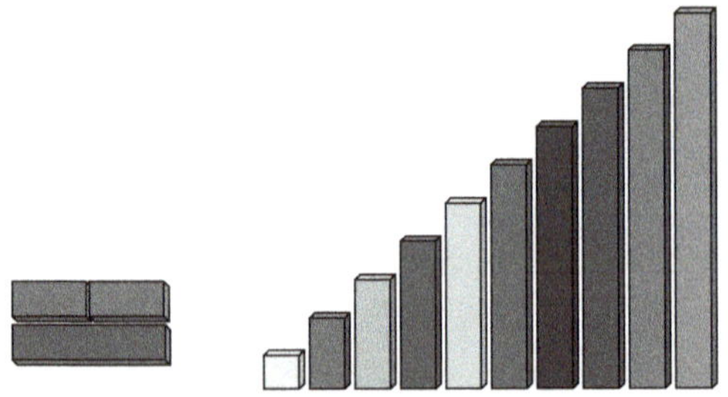

2. Ask, "What number of red units does the purple rod measure? Work with your group members to measure the purple rod using the red units. **Show Me** how you found your answer." **Observe** as students work and make note of students who can easily iterate units as they measure the length of the purple rods. Look for students who use one red rod multiple times and students who line up multiple red rods to measure the length of the purple rod. It may be necessary to ask questions to help students remember to line up the rods end to end and to line up the units end to end with no gaps or overlaps. Students should share that the purple rod is two red units long.

3. Allow students to measure the length of the dark green, brown, and orange rods using red rods as the unit. **Observe** students as they work and make note of students who can easily iterate units as they measure the length of the other rods. Do students successfully use the rules for measuring length (lining up objects end to end without any overlaps or gaps)? You may find it useful to use these data in determining student grouping and scaffolds for the task.

4. Allow time for students to share their findings.

> **! PRODUCTIVE STRUGGLE**
>
> Students may share two different measurements for one of the rods. Share both measurements with the class and ask students to tell how they got each measurement. Which one is correct? This is a great opportunity to allow students to engage in student-to-student discourse, construct viable arguments, and critique the reasoning of others.

Note: Consider using the Observation and Show Me tools for recording responses discussed above (see Appendix B).

FACILITATE

1. Say, "Now that we have measured lengths, we're ready to complete a measuring task. How many people remember the story of Goldilocks and the Three Bears?" It may be helpful to have a copy of the book available or a screenshot of an image from the traditional story to show to students. Show students the Goldilocks picture and say, "This is Goldilocks. Baby Bear is twice as tall as Goldilocks. Mama Bear is four times as tall as Goldilocks. Papa Bear is six times as tall as Goldilocks. Use the Goldilocks unit as your measuring tool as you determine the heights of Baby, Mama, and Papa Bear. Then make each member of the Bear family. How tall is each member of the Bear family? Be prepared to show me how you know."

2. Distribute the Make the Bear Family page, Goldilocks measuring units, construction paper, scissors, glue sticks, and pencils/markers to students. Allow students to work with their small groups. Circulate and observe students as they work, making note of students' strategies for using the Goldilocks unit to create the members of the Bear family. Ask, "How did you use the Goldilocks unit to determine the heights of the bears? How would you describe the height of Baby Bear in Goldilocks units? Mama Bear? Papa Bear? How did you use the rules for measuring?"

3. Continue with the task. Say, "Use the Goldilocks unit or the Baby Bear unit to make a tree for the forest where the bears live. Your tree has to be at least two Goldilocks units tall. How many Goldilocks units tall is your tree? How many Baby Bear units tall is your tree? Be prepared to show how you know how tall your tree is." Allow students to work with their groups to make their trees.

4. Allow student groups to join another group of students for a group-to-group share. Say, "Share your work. Look at their bears. What do you notice? Did they solve the task the same way you did? Did they solve it differently?" Allow time for students to share their completed work, explain their answers, and ask questions if needed within their new groups.

5. Have student groups trade trees with each other. Instruct students to measure the height of their classmates' tree with the Goldilocks and Baby Bear units. Say, "Compare your tree to your friends' tree. Which tree is taller? How many Goldilocks units taller? How many Baby Bear units taller?"

> **! PRODUCTIVE STRUGGLE**
>
> In order for students to have opportunities to make sense of tasks, teachers must resist the urge to intercede too soon. Allow time for students to make sense of the problem and use problem-solving strategies to determine how to approach the task. Use questioning to facilitate student understanding.

> **STRENGTHS SPOTTING**
>
> Providing opportunities for students to make conjectures and then collaborate with peers as they try to prove or disprove them fosters development of strengths in communication and reasoning and proof—two critical components of mathematical proficiency (NRC, 2001).

CLOSE: MAKE THE MATH VISIBLE

1. Bring the groups back together for a whole-group discussion. Ask, "How tall is each member of the Bear family in Goldilocks units? How do you know? Show me or tell me how you determined each of the lengths." Student work should show that Baby Bear is 2 Goldilocks units tall, Mama Bear is 4 Goldilocks units tall, and Papa Bear is 6 Goldilocks units tall. Their work should reflect these measurements. Remember: Young children exhibit varying abilities with fine motor tasks, which means that their depiction of the Bear family members may not be "exact"; however, appropriate depictions and measuring should show the proportional relationship between the heights of the members of the family.

2. Post the trees in the class forest for a center activity. Allow students to view and measure the completed trees.

Exit Task. Goldilocks Measures Up

Find three classroom objects and measure each object in Goldilocks units.

» One that is taller than Baby Bear

» Name of object _____

How tall is the object in Goldilocks units? _____

My object is _____ units taller than Goldilocks.

» One that is taller than Mama Bear

» Name of object _____

How tall is the object in Goldilocks units? _____

My object is _____ units taller than Goldilocks.

» One that is taller than Papa Bear

» Name of object _____

How tall is the object in Goldilocks units? _____

My object is _____ units taller than Goldilocks.

TASK 37: MAKE THE BEAR FAMILY STUDENT PAGES

online resources — To download printable resources for this task, visit **resources.corwin.com/ ClassroomReadyMath/K-1**

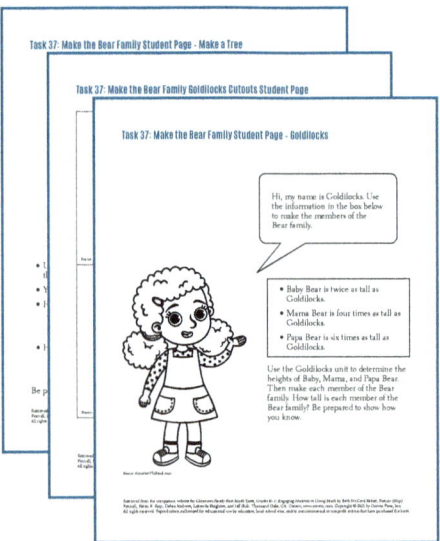

POST-TASK NOTES: REFLECTION & NEXT STEPS

Task 38
Comparing Cups

Describe and compare measurable attributes

TASK

Comparing Cups

Figure 12.3 Pitcher of Lemonade

The cost of lemonade at Field Day (Figure 12.3) depends on the amount the cup holds. You pay the least for the cup that holds the least lemonade and the most for the cup that holds the most lemonade. Maria is organizing the drink cups at the concession stand to make it easier to sell cups of lemonade. Can you help her put these cups in order from the one that holds the least to the one that holds the most? Which cup holds more? Which cup holds less? Show how you know.

TASK PREPARATION

- Prior to engaging in this task, it might be helpful if students have had prior experience with exploring capacity.

- Each group of students will need a set of cups in the sizes indicated. Place the 5, 10, 12, and 16-ounce cups in a baggie for easy distribution. See notes on labeling the cups in the Materials section. Decide which fill material students will use to measure capacity. Use the plastic dish pans or foil pans as trays to make clean up easier.

LAUNCH

1. Say, "What words can we use to describe how much one cup holds compared to another cup?" Students should respond that we can use the words *more* and *less* to describe how one cup compares to another. Show students the 8-ounce (cup B) and 9-ounce (cup C) cups. Ask, "What words could you use to describe these cups?

Mathematics Standard

- Directly compare two objects with a measurable attribute in common, to see which object has "more of"/"less of" the attribute, and describe the difference. *For example, directly compare the heights of two children and describe one child as taller/ shorter.*

Mathematical Practices

- Construct viable arguments and critique the reasoning of others.
- Model with mathematics.

Vocabulary

- capacity
- more
- less
- most
- least
- fill
- estimate
- fewer/fewest

Materials

- set of plastic cups in 5, 8, 9, 10, 12, and 16-ounce sizes (label the cups F, B, C, E, D, and A respectively). Select cups of varied dimensions such as tall and skinny versus short and wide so that it's not obvious which cup has the greatest capacity, one set per group of three students

- beans for filling cups (all groups will need access to fill materials)

- plastic dish pan or large disposable foil pan, one per group of students

- paper towels for clean up

- Comparing Cups student page

Turn and talk to your partner and tell how you could use beans to find out how much each cup holds. What would you tell someone about which of these cups would hold more or fewer beans?" Record students' responses.

2. Divide students into pairs and provide each pair with one 8-ounce and one 9-ounce cup. Tell students: "Examine the two cups. Which of these cups do you think will hold the most beans? Why?" Distribute beans and a dish pan to student pairs. Tell students: "Work with your partner to find out which cup holds the most beans. Put the cups in order from the one that holds the fewest beans to the one that holds the most." **Observe** students as they fill each cup with beans. What methods are students using to measure the capacity of each cup?

3. Ask, "Does cup B hold more or fewer beans than cup C? **Show Me** how you know. Is this different from your earlier estimate?"

Note: Consider using the Observation and Show Me tools for recording student responses (see Appendix B).

FACILITATE

1. Say, "We just found out that cup C holds more beans than cup B. Here is a baggie with four more cups. What can you tell me about these cups?" Allow time for students to examine the other cups. Have students make estimates about which cups hold the most and fewest beans and record student predictions on chart paper. Say, "We're going to use these cups and our beans to solve a problem." Share the task with students.

"The cost of lemonade at Field Day depends on the amount the cup holds. You pay the least for the cup that holds the least lemonade and the most for the cup that holds the most lemonade. Maria is organizing the drink cups at the concession stand to make it easier to sell cups of lemonade. Can you help her put these cups in order from the one that holds the least to the one that holds the most? Which cup holds more? Which cup holds less? Show how you know."

2. **Observe** students as they work. What methods are students using to compare

! PRODUCTIVE STRUGGLE

It's important to provide all students with the opportunity to problem solve and discover strategies for comparing the cup capacities. However, if students struggle with finding a starting point on this task, you can scaffold students' work by focusing their work on comparing two cups to then introduce a new cup. Students can use what they know about the two cups to order the new cup. Continue with this procedure until students have compared and ordered all six cups.

the cups? Do students pour the beans from a smaller cup into a larger cup to determine which one holds more? Do students fill each cup and count the number of beans in each cup? Are students able to place cup B and cup C in the correct place? Possible questions to ask as you visit student pairs are as follows: "Which cup holds more? Which cup holds less? Where do the other cups belong? How do you know?"

3. Tell students, "When you're finished, line up your cups with the cup that holds the fewest beans first and the one that holds the most beans last. Place the other cups where they belong in the order. If you would like to, you can draw a picture to show your answer."

4. **Hinge Question.** How did you use what you know about the first two cups we worked with, cup B and cup C (hold B and C cups up), to place all of the cups in order?

5. Allow each pair of students to join another pair for a *Pair-to-Pair Share*. Allow time for students to share their completed work, explain their answers, and ask questions.

STRENGTHS SPOTTING

Highlight students' thinking as they create arguments to support their reasoning. In this case, students may explain that the information they learned about the capacity of cups B and C helped them figure out the capacity of the other cups. Recognize that students' explanations may be tentative or even awkward as they share emerging ideas. This type of sharing builds content strengths.

CLOSE: MAKE THE MATH VISIBLE

1. Allow students to participate in a Gallery Walk where they visit each pair to see their results. This is a good time for students to revisit their estimates from the beginning (Facilitate portion) of the task.

2. Bring students together for a whole group discussion. Allow students to discuss the results of their comparing task. Students should report that the order of the cups from most to least is A, D, E, C, B, F. During the debriefing discussion, encourage students to explain strategies they used to compare the cups.

3. **Exit Task.** Maria can have free refills of lemonade if she brings her own cup to Field Day. Which cup should she bring, cup A, cup B, cup C, or cup E? Explain your answer.

TASK 38: COMPARING CUPS STUDENT PAGE

 To download printable resources for this task, visit **resources.corwin.com/ ClassroomReadyMath/K-1**

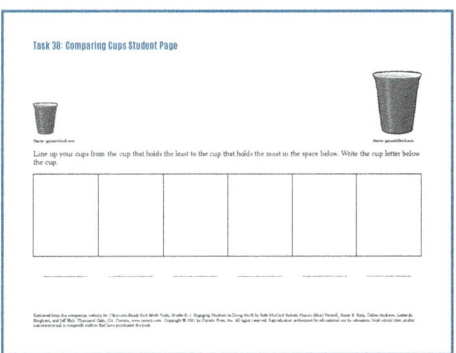

Task 39
Guess My Sorting Rule

Classify objects and count the number of objects in each category

TASK

Guess My Sorting Rule

We have been sorting all types of objects based on many different rules. Today we are going to play a guessing game. You and your partner have a set of buttons. Sort your buttons into two different groups, but do not tell anyone the rule you used to sort them. Be creative. Another pair will try to guess the rule you used to sort your buttons.

TASK PREPARATION

- Students should have engaged in multiple opportunities to sort and classify a variety of objects prior to engaging in this task.

- Students will need to work in pairs for this task. Prepare baggies of buttons ahead of time for easy distribution to students.

LAUNCH

1. Display the Which One Doesn't Belong? prompt (Figure 12.4).

Figure 12.4 Which One Doesn't Belong?

Source: Buttons by pixabay.com and redcollegiya/iStock.com

Mathematics Standard

- Classify objects into given categories based on their attributes; count the numbers of objects in each category and sort the categories by count.

Mathematical Practices

- Make sense of problems and persevere in solving them.

- Model with mathematics.

Vocabulary

- sort
- attributes
- category
- rule

Materials

- multiple sets of buttons, one set per pair of students. Try to find a variety of types of buttons to encourage sorting based on a variety of categories. If real buttons aren't available for this task, you could use attribute buttons (a commercially available math manipulative).

- paper plates or sorting mats for sorting

- Guess My Sorting Rule student page, one for each pair of students

2. Say, "Look at the display. Which one doesn't belong? You have about two minutes of individual time before working with a partner. Be prepared to explain which one doesn't belong and tell why." After about two minutes, allow students to share their thinking with a partner. Encourage students to use a variety of methods such as math reasoning, pictures, or words to prove that their decision makes sense.

3. **Observe** and monitor students' sharing. You may want to ask questions that prompt students to explain why their decision about which one doesn't belong makes sense.

4. Allow students to share their responses and reasoning. Encourage students to explain their reasoning for why one of the choices doesn't belong. Record student responses.

5. **Observe** as students discuss the attributes of the buttons depicted, including color, size, shape, and number of holes. Make a list of the attributes that describe the buttons in the Which One Doesn't Belong? task.

Note: Consider using the Observation tool for recording the attributes observed (see Appendix B).

6. If available, read the short story "A Lost Button" from the book *Frog and Toad Are Friends* by Arnold Lobel or *Button Box* by Margarette Reid to students. Discuss the story details and ask students to describe the attributes of the buttons in the story (color, size, shape, number of holes, material, and thickness). Ask, "Are there any additional attributes we need to add to our list?" Record student responses.

FACILITATE

1. Regroup the students into pairs and distribute a button collection to each pair. Allow time for students to examine the buttons in their collections and talk about the different types of buttons.

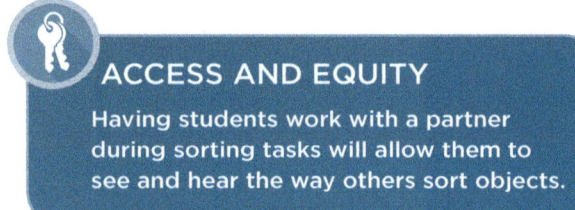

ACCESS AND EQUITY

Having students work with a partner during sorting tasks will allow them to see and hear the way others sort objects.

2. Distribute paper plates or sorting mats to each pair. Say,

"Today we are going to play a guessing game. You and your partner have a set of buttons. Sort your buttons into two different groups, but do not tell anyone the rule you used to sort them. You and your partner should both know what secret rule you used to sort the buttons. Another pair will look at your sorted buttons and try to guess what rule you used to sort them. Be creative."

Allow students to sort their buttons.

3. **Observe** the student pairs as they sort their buttons, asking them to quietly describe how they sorted the buttons.

4. Then, **Interview** the pairs, asking them to tell you the name of each group. Ask, "Why are these buttons grouped together? What is the name of this sorting group? What rule did you use to sort your buttons? How many buttons are in each of your sorting groups? Do any of your groups have the same amount? How could you prove they are the same number?" Encourage students to use the recording sheet to record their sorting rules and categories.

PRODUCTIVE STRUGGLE

Allow time for students to examine the objects and select sorting categories. If students have difficulty getting started, you may prompt students by asking how they plan to sort the buttons. If needed, suggest a sorting category, or have students use one of the attributes on the list generated at the beginning of the activity.

Note: Consider using the Observation and Interview (small group) tools for recording student responses (see Appendix B).

5. Allow each pair of students to join another pair for a *Pair-to-Pair Share*. Ask student pairs to take turns trying to guess the sorting rule for their assigned pair. Encourage the students to discuss the sorting rules, discuss their observations, and ask questions if desired. Rotate among the groups and observe the interaction between students as they try to guess each other's sorting rules. Ask questions such as, "What's the same about how you sorted your buttons? What is different about how you sorted your buttons? What clues did you use to determine how your friends sorted their buttons?"

6. Have partners return to their working space and instruct students to complete a second sort using a different attribute. After the new sort, have students join another group for a *Pair-to-Pair Share*. Allow students to sort their buttons a third time and engage in a *Pair-to-Pair Share*.

! PRODUCTIVE STRUGGLE

Have a few extra buttons available as you visit student pairs. If a pair is struggling to figure out the rule, give them a button and ask them to predict which group the new button would fit into.

! PRODUCTIVE STRUGGLE

Students may think that objects can only be sorted one way based on one attribute. By encouraging students to sort the same set of objects using different attributes each time, students develop an understanding that objects can be sorted by more than one attribute.

CLOSE: MAKE THE MATH VISIBLE

1. Bring the students together for a whole-group discussion. Allow students to share the following with the class. Ask, "What attributes did you use to create a rule for the sort? Which group had the most buttons? Which group had the fewest buttons? Did anyone sort their buttons in the same way? Did anyone have a rule that was hard to guess? If yes, why was it more difficult to guess than the others?" If needed, add any new attributes to the class list introduced in the task's launch.

2. **Hinge Question.** Sherry sorted some shapes. Her work is shown in Figure 12.5. What shape could you add to the groups in her sort? Why would you add that shape?

Figure 12.5 Sherry's Shape Sorting

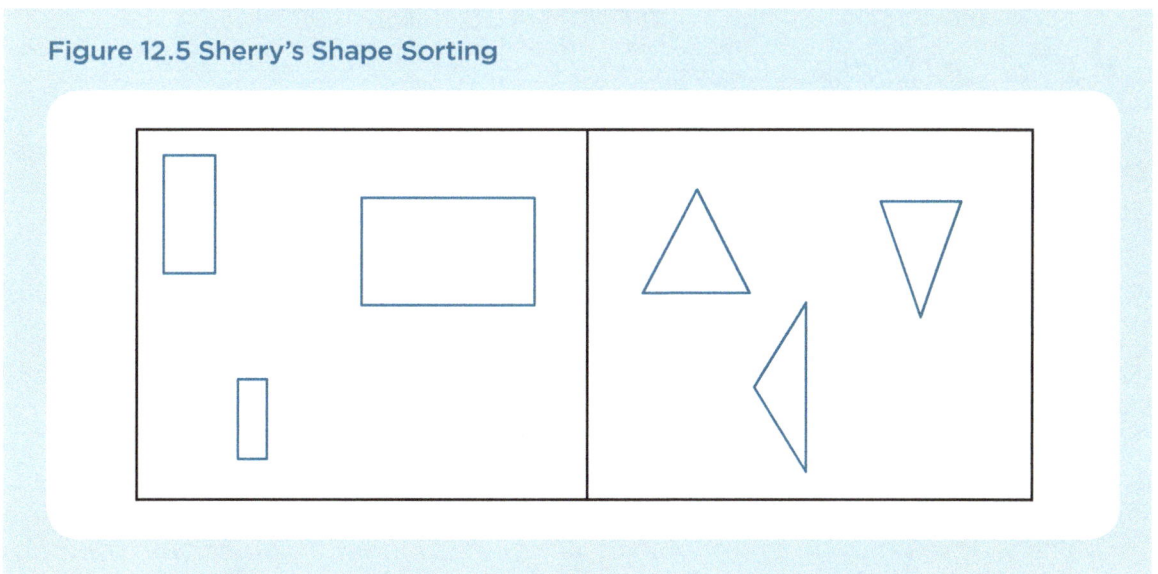

TASK 39: GUESS MY SORTING RULE STUDENT PAGES

 To download printable resources for this task, visit **resources.corwin.com/ ClassroomReadyMath/K-1**

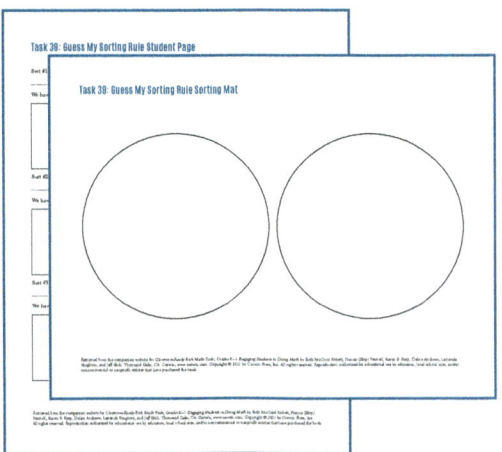

POST-TASK NOTES: REFLECTION & NEXT STEPS

_____

_____

_____

_____

_____

_____

_____

_____

Task 40
Which Worm Is Which?

Measuring lengths indirectly and by iterating length units

TASK

Which Worm Is Which?

Figure 12.6 Three Worms

Source: eliflamra/iStock.com

You have three worms, but the teacher forgot to color them (see Figure 12.6). Use what you know about putting objects in order by length to figure out which worm to color red, which worm to color yellow, and which worm to color blue.

• Robert Red worm is longer than Brianna Blue worm.

• Yolanda Yellow worm is shorter than Brianna Blue worm.

• Which worm is which? Color the worms to show your answer.

TASK PREPARATION

• Students should have prior experience with putting objects in order from shortest to longest length before engaging in this task.

• Prepare baggies of classroom objects; sets of red, yellow, and blue crayons; and copies of Worm Pictures ahead of time.

LAUNCH

1. Consider reading a children's literature book about comparing objects by size to students (e.g., *The Best Bug Parade* by Stuart J. Murphy).

Mathematics Standard

• Order three objects by length; compare the lengths of two objects indirectly by using a third object.

Mathematical Practices

• Make sense of problems and persevere in solving them.

• Attend to precision.

Vocabulary

• length
• longer, longest
• shorter, shortest
• compare
• order

Materials

• baggie of classroom objects such as crayons, markers, and unsharpened pencils, one per pair of students

• Worm Pictures student page, one per pair of students

• scissors, one per pair of students

• a set of red, yellow, and blue crayons, one per pair of students

• large piece of construction paper, one per pair of students

• glue stick, one per pair of students

• color tiles

• connecting cubes

» If a book is read, allow time for the students to discuss the measurement words used in the story such as ordering the bugs by length.

» If a book is not read, provide representations of objects of different lengths (e.g., cards, drawings of bugs, rods of connecting cubes) and discuss the lengths of objects and comparing and ordering their lengths.

2. Divide students in pairs and have each pair take the crayon and unsharpened pencil out of their baggie. Tell students: "Turn and talk to your partner and show them which object is longer. How do you know it is longer?"

3. **Observe** students as they work to see if students line up the crayon and pencil at the endpoints to compare the lengths.

4. **Show Me.** Give each student pair the marker and ask them to place three objects in order from shortest to longest. Ask, "How do you know which object is the longest? Shortest? Show me how you know."

5. Give each student pair a set of Worm Pictures (student page) and scissors. Allow time for students to cut apart the worm cards on the heavy dark line. Once students have cut apart the cards, ask students to place the worms in order from shortest to longest.

6. Observe students as they work to see if the students compare the worms to find the longest and shortest. Do they understand that the third worm should be between the shortest and longest worms? Ask, "How did you figure out which worm was the shortest and which was the longest? How did this help you find out which worm belongs in the middle?"

Note: Consider using the Observation and Show Me tools for recording responses (see Appendix B).

ACCESS AND EQUITY

As you ask questions, listen with curiosity! Challenge the tendency to listen for errors or look for opportunities to correct thinking. Listening with curiosity means listening to each student with the goal of really seeing the task from their point of view. Such a perspective can often open your eyes to strong reasoning and important mathematical ideas that may not be evidenced in traditional ways.

FACILITATE

1. Tell students, "Now we're going to use what we know about putting objects in order from shortest to longest length to solve a math riddle."

2. Read the task to students: "You have three worms, but the teacher forgot to color them. Your task is to work with your partner to determine what color each worm should be. Use what you know about putting objects in order by length to figure out which worm to color red, which worm to color yellow, and which worm to color blue. Here are the clues: Robert Red worm is longer than Brianna Blue worm. Yolanda Yellow worm is shorter than Brianna Blue worm. Which worm is which? Color the worms to show your answer."

3. **Observe** student pairs as they work. Make note of the strategies the students use to compare the worms' lengths. Do they line up the endpoints? How do they use the clues to determine what color to make each worm?

PRODUCTIVE STRUGGLE

Sometimes it can be tempting to step in and "rescue" students when they appear to struggle with a critical thinking task. However, in order to encourage students to do the sense making, encourage them to work with their peers on this task and allow time for them to use the clues to make sense of the task. Encourage them to use what they know about the worms to identify the color of one worm at a time.

Do they use what they know about the length of two of the worms to decide which worm is the longest in length and shortest in length, and which worm is in the middle or between those lengths?

4. Provide student pairs with a large piece of construction paper and a glue stick to glue the Worm Pictures in order from shortest to longest length.

5. Allow each pair of students to join another pair for a *Pair-to-Pair Share*. Allow time for students to share their completed work, explain their answers, and ask questions.

CLOSE: MAKE THE MATH VISIBLE

1. Conduct a whole-group discussion. Reread the riddle to students: "Robert Red worm is longer than Brianna Blue worm. Yolanda Yellow worm is shorter than Brianna Blue worm. Which worm is which? This is the riddle we started our lesson with. We knew everything we needed to know about each worm's length, but you had to provide the missing information to tell which worm is which." Ask, "Based on your work with your partner, which worm is red, yellow, or blue?" Students should share that the longest worm is red, the shortest worm is yellow, and the worm in the middle or between those lengths is blue. Say, "Explain how you know. How did the clues help you figure out which worm is which?"

2. **Exit Task.** Provide students with paper for drawing, connecting cubes, or counting tiles. Then ask them to make a drawing or tower to show the result of the measurement riddle below.

Ayesha's tower is taller than Jose's tower.

Ayesha's tower is shorter than Maggie's tower.

Maggie's tower is _____ Jose's tower.

Share your drawing or towers with others.

TASK 40: WHICH WORM IS WHICH? STUDENT PAGE

 To download printable resources for this task, visit **resources.corwin.com/ ClassroomReadyMath/K-1**

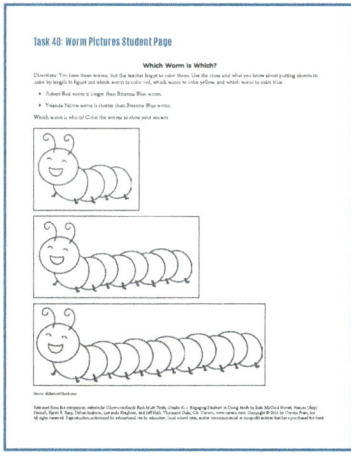

Task 41
Measure and Compare

Measure lengths indirectly and by iterating length units

TASK

Measure and Compare

Figure 12.7 Objects for Measuring

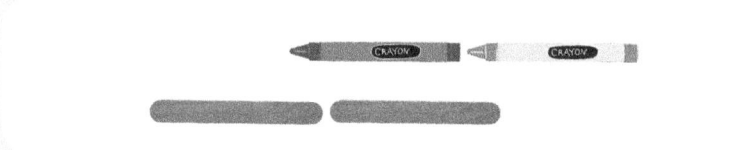

Source: Crayons by pixabay.com; Popsicle stick by publicdomainpictures.net

You and your partner will measure the length of each object in the measurement bag (Figure 12.7) using two different measurement units, and compare the results. Is one measurement of the object greater than the other? Explain why the two measurements of the same object are different.

TASK PREPARATION

- Students should have multiple opportunities to measure objects using a variety of measurement units (standard and nonstandard) prior to engaging in this task.

- Prepare measurement bags ahead of time.

LAUNCH

1. Bring students together for a whole-group discussion. Make up your own story or, if possible, locate and read a children's literature book about measuring length, such as *Twelve Snails to One Lizard* by Susan Hightower. Discuss story details with the students. For *Twelve Snails to One Lizard*, ask questions such as, "Milo measured the length of the dam using snails, lizards, and Betty Jo Boa. What happened to the number of animals needed as Milo used a different animal to measure the length? Why? Turn and talk to your partner and share your thinking."

ACCESS AND EQUITY

Partner talk before whole-class sharing helps to ensure that every student has an opportunity to think before the whole-group discussion and gives the teacher an opportunity to circulate and listen for two or three ideas to highlight with the whole group. Students should be allowed to share their own idea or an idea they heard from their group.

Mathematics Standard

- Express the length of an object as a whole number of length units, by laying multiple copies of a shorter object (the length unit) end to end; understand that the length measurement of an object is the number of same-size length units that span it with no gaps or overlaps. *Limit to contexts where the object being measured is spanned by a whole number of length units with no gaps or overlaps.*

Mathematical Practices

- Attend to precision.
- Use appropriate tools strategically.

Vocabulary

- measure
- length
- estimate
- unit

Materials

- large and small paper clips, one baggie of each for each pair of students

- a variety of nonstandard units of varying lengths, such as color tiles, centimeter cubes, connecting cubes, rigatoni pasta pieces, pretzel sticks, teddy bear counters, and beans (You'll need a large enough quantity of each item so that students can choose two different measurement units for the measurement task.)

- measurement bags: identical baggies of classroom objects such as marker, crayon, pencil, glue stick, toy car, one per pair of students

- pencils

- Measure and Compare student page, one per pair

Note: If the story telling or children's literature activity is not completed, begin the task lesson Launch at step 3.

2. Allow time for student pairs to share their thinking. Record student ideas on chart paper (Figure 12.8). Students should share that Milo needed more (36) snails to measure the length of the dam than the other units (lizards and the snake) because the snails were a shorter length.

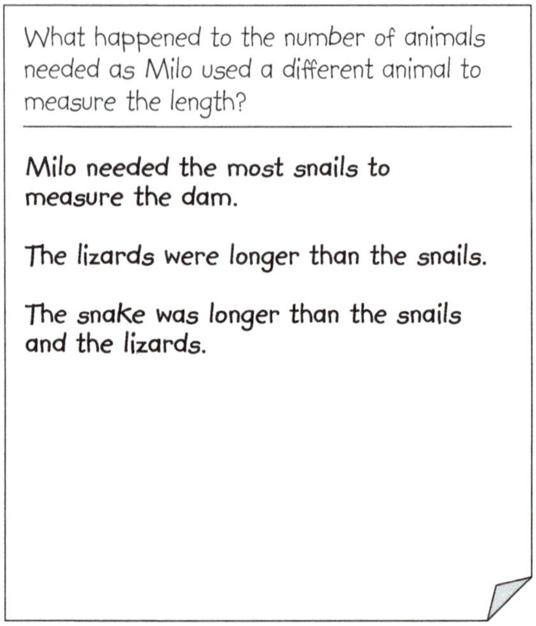

Figure 12.8 Sharing Student Ideas

What happened to the number of animals needed as Milo used a different animal to measure the length?

Milo needed the most snails to measure the dam.

The lizards were longer than the snails.

The snake was longer than the snails and the lizards.

3. Show students a stapler. Say, "We've been learning a lot about measuring length using a variety of units. If we were to measure the length of this stapler, what are some things we need to know? What are some rules we use for measuring the length of an object?" Allow time for students to respond and remind each other of the "rules" for measuring the length of an object: estimate the number of units needed to measure the object, and line up units end to end with the object with no gaps or overlaps. Ask, "How do we describe the length of an object?" (Students should remind the group that they need to use a number and a unit to describe the length of an object.)

4. Select a student volunteer to estimate and then measure the length of the stapler with large paper clips (Figure 12.9). Record the measurement. Ask, "If we measure the length of this stapler using small paper clips, what do you think will happen to the number of units it will take to measure the length of the stapler? Why?"

Figure 12.9 Stapler Measured With Large Paper Clips

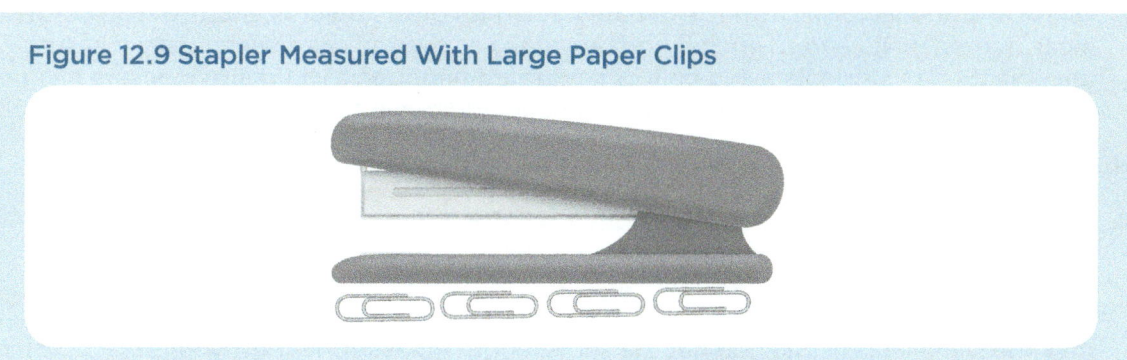

Source: ONYXprj/iStock.com

5. Then select a different student volunteer to estimate and then measure the length of the stapler using small paper clips (Figure 12.10). Record the measurement. Ask, "What did you notice? What do you wonder about these two different measurements of the same object?"

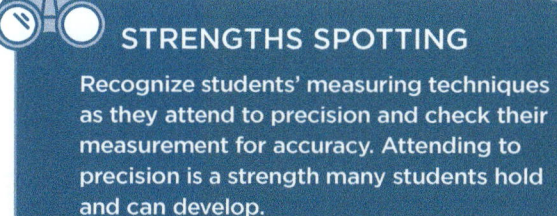

STRENGTHS SPOTTING

Recognize students' measuring techniques as they attend to precision and check their measurement for accuracy. Attending to precision is a strength many students hold and can develop.

Figure 12.10 Stapler Measured With Small Paper Clips

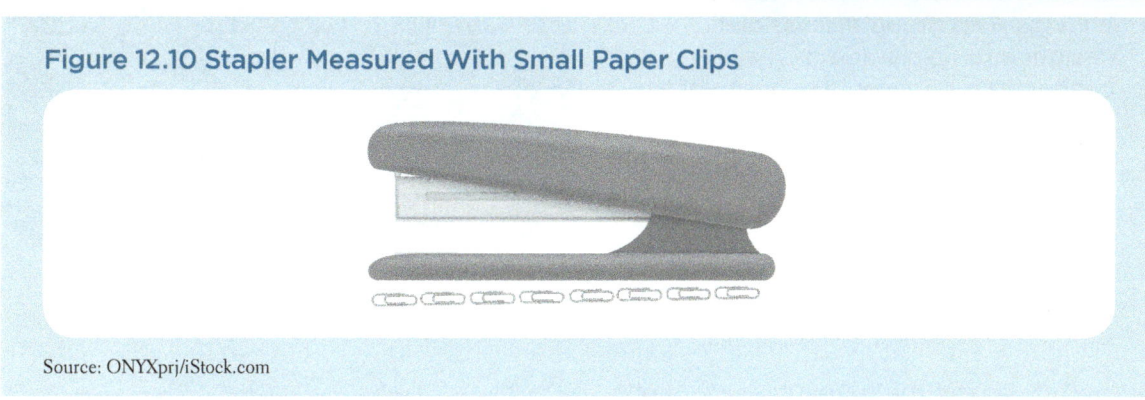

Source: ONYXprj/iStock.com

6. Have the students turn and talk with a partner and then share their ideas. Record student responses. If students need additional scaffolding, ask questions such as, "Which unit—large paper clip or small paper clip—is longer? Which measurement is greater?" to prompt student thinking.

FACILITATE

1. Organize the students in pairs. Tell students: "We measured the stapler using two different measurement units and compared our results. You and your partner will measure each object in this measurement bag using two different measurement units and compare the results. Before you measure, estimate which unit will produce a measurement that is greater than the other."

2. Distribute a measurement bag, recording sheet, and two different nonstandard measurement units to each pair of students. Make sure that each pair of students receives measurement units of two different lengths.

ACCESS AND EQUITY

Consider allowing students to choose their measurement units. To ensure that they select different-sized units, offer them a choice from measurement units in pile A and measurement units in pile B. Providing choice promotes student agency and positions them as capable and competent learners.

3. **Observe** students to see if they estimate the length of the objects and then line up the measurement units end to end without any gaps or overlaps as they actually measure the objects. Are students using only one measurement unit at a time to measure each object? Make a note of students who may need additional support.

Note: Consider using the Observation tool for this activity (see Appendix B).

4. Allow each pair of students to join another pair for a *Pair-to-Pair Share*. Allow time for students to share their measurement results, ask questions, and share what they noticed. Possible questions students may ask each other are as follows: "Which two measurement units did you use? How are your two measurement units different? Are they the same length? Did you get the same length measurement for your objects using both different units? If no, how are they different?" Tell students: "After meeting with your new group members, record any new things you noticed on your recording sheet."

CLOSE: MAKE THE MATH VISIBLE

1. Bring students to a whole-group discussion. Tell students: "We used two different measurement units to estimate and measure the length of classroom objects."

2. Ask, "What did you notice about the measurements with the different units?" Allow students to share their findings. This group discussion is designed to help students make connections between the length of the unit and the number of units it takes to measure an object's length.

ACCESS AND EQUITY

Providing opportunities for students to reason, test, and build on their prior ideas about how to measure and use those measurements to compare will support students to make connections about length, number of units, and size of the measurement unit.

3. Ask the following questions to guide the discussion related to measuring with the two measurement units and completion of the Measure and Compare recording sheet:

   » Did you get the same measurement for each object using both measurement units?

STRENGTHS SPOTTING

Students who are willing to take risks should be celebrated. Estimating *before* measuring can help students notice and begin to prevent accuracy errors. Encourage students to ask, "What kind of a result do I *expect* to get? And what surprised me about the measurement?"

   » Were both measurement units the same size?

   » Which length measurement was greater? (for each of the objects measured).

   » Did anything surprise you about your measurement results?

   » What can you say about the relationship between the size of the unit and the number of units it takes to measure the length of an object? How do you know?

4. Say, "Turn and talk to your partner and share your thinking." Record student responses.

5. **Exit Task.** Show students the stapler from the beginning of the task and remind them how long the stapler was in large paper clips. Give each student a centimeter cube. "Use what you know about the size of a unit and the number of units it takes to measure an object to tell me if you think we will need more, fewer, or the same amount of centimeter cubes as large paper clips to measure the length of the stapler. Explain your answer using words, pictures, or both."

TASK 41: MEASURE AND COMPARE STUDENT PAGE

Task 41: Measure and Compare Student Page

Directions: Using one measurement unit at a time, measure the length of each object. Record the measurements on the recording sheet.

Classroom Object	Length Using _____	Length Using _____
glue stick		
marker		
pencil		
crayon		
toy car		

We noticed . . .

After the Pair-to-Pair Sharing, we noticed . . .

POST-TASK NOTES: REFLECTION & NEXT STEPS

Measurement

Time

TASK 42: GRADE 1: SOCCER TIME

Tell and write time in hours and half-hours using analog and digital clocks.

TASK 43: GRADE 1: WHAT TIME IS IT?

Tell and write time in hours and half-hours using analog and digital clocks.

TASK 44: GRADE 1: FUN FRIDAY SCHEDULE

Tell and write time in hours and half-hours using analog and digital clocks.

Anticipating Student Thinking: Time is a mathematics application that is typically a standard within the measurement content strand. However, telling time, estimating times, considering elapsed time, and determining the differences in times are just a few of the ways in which adults use time every day. Students at the preschool level actually begin to learn about time informally as they recognize the difference between night and day, and mornings, afternoons, and evenings. They also may, more formally, "learn" about the specific time to go to bed, watch a television show, and so on. Formal instruction related to telling time begins at the first-grade level. All three tasks in this chapter focus on telling time to the nearest hour and half hour, with task lesson activities involving analog and digital time and writing these timed amounts. Also note that two of the tasks informally introduce the fraction $\frac{1}{2}$ as related to time to the half hour. As you plan for implementation of the unit's tasks, recognize that it's likely that your students will have a range of experiences related to thinking about and using time. Your use of Observation and Show Me in the tasks should serve as valuable strengths-spotting opportunities to assess such differences.

THINK ABOUT IT

Extend the chapter's task activities to include at-home opportunities for your students to ask family members about time and their use of time. For example, when do they get up in the morning? When do they go to bed at night? Do they use digital or analog watches and clocks? Or, how about asking a family member to name one thing that takes about an hour and one thing that takes about half an hour? Connecting these school-based task lessons to home and family use strengthens the beginnings of this lifelong mathematics application.

Mathematics Standard

- Tell and write time in hours and half-hours using analog and digital clocks.

Mathematical Practices

- Construct viable arguments and critique the reasoning of others.
- Use appropriate tools strategically.

Vocabulary

- time
- hour
- o'clock
- analog clock
- hand
- face

Materials

- student mini analog clocks with moveable hands, one per student, or use a blank student recording sheet
- Soccer Time student page, one per student
- pencils
- demonstration analog clock

Task 42
Soccer Time

Distinguish between the hour hand and the minute hand

TASK

Soccer Time

Soccer practice usually starts at 1:00 p.m. Samantha arrived at practice when her clock showed the time shown in Figure 13.1.

Figure 13.1 Samantha's Arrival Time

Martin arrived at practice when his clock showed the time shown in Figure 13.2.

Figure 13.2 Martin's Arrival Time

When he arrived, he was the only one at practice. He told Samantha that everyone else was late for practice. Do you agree or disagree with Martin? Show how you know.

ALTERNATE LEARNING ENVIRONMENT

To facilitate this task via distance learning, consider a digital clock such as the Math Clock app from the Math Learning Center (mathlearningcenter.org/math-clock/).

TASK PREPARATION

- Students will need access to the mini clocks throughout this task. It may be helpful to package the clocks in baggies for each table group to help with distribution for the Launch activity.

- To be successful on this task, students should have prior experience with tasks where they learned the meaning of the hour hand and minute hand and how to use the hour hand to tell time to the hour.

- As students work on this task, you may observe some students who still have some incomplete conceptions about the purpose of the hour hand and telling time to the hour. Be prepared to support these students with additional scaffolding and review activities on using the hour hand to tell time to the hour.

- Consider sharing a children's literature book that features telling time on the hour, such as *The Grouchy Ladybug* by Eric Carle or *Time to …* by Bruce McMillan, with the students.

LAUNCH

1. Ask, "When you hear the word *time*, what do you think about? When do you talk about or use time at home?" Distribute mini-analog clocks to students. If *The Grouchy Ladybug* is available, ask students to show the times described in the story on their clocks as you read the story. If you do not have a time-related book, ask the students to suggest an activity for each of the following times and then discuss the actual time for each event. Then have the students show each of the times on their analog clock.

   » What time does our class have lunch?

   » What time do we go to art class?

   » When does school end for the day?

   » What time do you come into our class each day to start the day?

2. Using either scenario, observe students as they show the times in the story or the school day on their clocks. Make note of students who have difficulty showing the time to the hour or who have trouble distinguishing the hour hand from the minute hand.

3. Show students the demonstration clock and ask students to show you the hour hand on their mini-clocks. Ask, "How is the hour hand different from the minute hand?"

4. Allow time for students to respond. Say, "Turn and talk to a partner and tell them some things you know about using the hour hand to tell time." Look for these possible responses:

   » The hour hand points to the hour.

   » The hour hand points to each hour as it moves from 1 to 12 in a circle around the clock face.

   » When the hour hand points to a number on the clock like 3, the time is three o'clock.

   » The hour hand points directly at a number when the time is on the hour.

   » When the hour hand is exactly halfway between two numbers on the clock face, the time is half past the hour.

   » When the hour hand moves all the way around the clock from one number back to the same number, twelve hours have passed.

5. Record student responses on chart paper. Ask, "How did you use this information to help you show the times we discussed?" Allow time for students to share their responses.

FACILITATE

1. Say, "Today I want you to use what you know about telling time to the hour to solve a task. Today's task is about two children who have soccer practice."

2. Distribute the Soccer Time student page to students. Read the task to students: "Soccer practice usually starts at 1:00 p.m. Samantha arrived at practice when her clock showed this time (see Figure 13.1). Martin arrived at practice when his clock showed this time (see Figure 13.2). Look at both clocks. What do you notice? What do you wonder? Turn and talk to a partner."

3. Allow students to share their responses. Students may point out that the hour hand on Samantha's clock is on 1, the hour hand on Martin's clock is on 12. The minute hand on Martin's clock is on the 1. Samantha's clock shows 1 o'clock. Martin's clock shows some time after 12 o'clock. (First-grade students typically are expected to tell time to the hour and half hour only, but some students may say that Martin's clock shows 12:05.)

4. Read the next part of the task: "When he arrived, he was the only one at practice. He told Samantha that everyone else was late for practice. Do you agree or disagree with Martin? Show how you know." Students should still have access to the mini-clocks. Allow time for students to solve the task individually. Circulate as students work and **Observe** their work. What types of strategies are students using? Do students use the mini-clocks to model 1:00 p.m.? Ask students to model Martin's time.

5. Place students in pairs. Allow students to share their work with each other.

CLOSE: MAKE THE MATH VISIBLE

1. Bring class together for whole-group discussion. Ask students, "Do you agree or disagree with Martin? Why?" Students should share that Samantha correctly arrived at soccer practice at 1:00 p.m., which was the time practice was supposed to start. Student justification for their choice should include information about the placement of the hour and minute hands on Samantha's clock and Martin's clock and why Samantha's clock tells the correct time and Martin's does not.

2. Select students to share their responses. Questions to use to guide student discussion could include the following: "How are the two clocks different? How can you use your clock to show how you know that Samantha wasn't really late for her soccer practice? Why might Martin have thought it was 1:00 when he looked at this clock?" (Show students a picture of Martin's clock.)

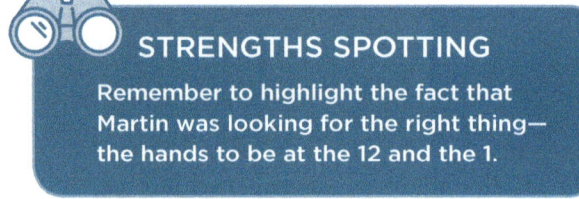

STRENGTHS SPOTTING

Remember to highlight the fact that Martin was looking for the right thing—the hands to be at the 12 and the 1.

3. Allow students to write or dictate advice for Martin on how to tell the difference between 12:05 and 1:00.

4. Refer to the student ideas about how they could use the hour hand to tell time.

5. **Hinge Question.** How did you use what you knew about the hour hand to help you on this task? Allow time for students to share their thinking.

6. Success on this task is dependent on students being comfortable with using the hour hand to tell time to the hour and being able to distinguish between the hour and minute hand. Using a one-handed clock that only features an hour hand is a great tool to help students get a better understanding of the meaning of the hour hand. You can use mini-paper plates, paper fasteners, and an arrow cut out to make one-handed clocks for students. Use a large paper plate to make a teacher demo clock, or purchase an

inexpensive analog clock from a discount store and remove the minute hand. Provide students with practice pointing the hand of the clock directly to designated numbers and identifying the time each time. Next, allow time for students to work with a partner and practice showing additional times to the hour. Make certain to include time for students to write or tell what they know about using the hour hand on a clock to tell time.

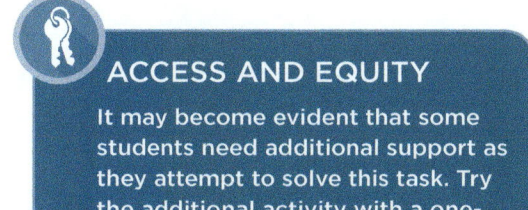

ACCESS AND EQUITY

It may become evident that some students need additional support as they attempt to solve this task. Try the additional activity with a one-handed clock to support student understanding.

TASK 42: SOCCER TIME STUDENT PAGES

 To download printable resources for this task, visit **resources.corwin.com/ ClassroomReadyMath/K-1**

POST-TASK NOTES: REFLECTION & NEXT STEPS

Mathematics Standard

- Tell and write time in hours and half-hours using analog and digital clocks.

Mathematical Practices

- Use appropriate tools strategically.
- Construct viable arguments and critique the reasoning of others.

Vocabulary

- hour
- minutes
- o'clock
- thirty
- half past
- hand
- face

Materials

- student analog clocks with moveable hands, one per student
- pencils
- demonstration analog clock
- What Time Is It? student page
- Blank Clocks and Hands student page (optional)
- circular number line (optional)
- whiteboards (optional)

Task 43
What Time Is It?

Tell and write time to the half hour

TASK

What Time Is It?

Zaara showed this time (Figure 13.3) on her clock.

Figure 13.3 Zaara's Clock

Gabriel said that the clock shows the time 10:30. Do you agree or disagree with Gabriel? Why? Show how you know.

TASK PREPARATION

- Prepare student materials ahead of time. If possible, place student clocks in baggies for easy distribution. Make plans for placing students in heterogeneous pairs.

ALTERNATE LEARNING ENVIRONMENT

To facilitate this task via distance learning, consider a digital clock such as the Math Clock app from the Math Learning Center: apps .mathlearningcenter.org/ math-clock/

LAUNCH

1. Have the students use their analog clocks to represent the following times to the hour: 2 o'clock; 10 o'clock; 8 o'clock.

2. **Show Me.** If we have to go to the library at half an hour after 2 o'clock, what would that look like on the clock? Use your analog clocks to set the time that is half an hour past 2 o'clock.

3. Represent and then discuss the times for 2 o'clock and two-thirty using analog and digital time.

4. Display the Which One Doesn't Belong? prompt (Figure 13.4). (Make paper mini-clocks and whiteboards available for student use.) Say, "Look at the display. Which one doesn't belong? You have a few minutes of individual time before working with a partner. Be prepared to explain which one doesn't belong and why." After about five minutes, allow students to share their thinking with a partner. As you **Observe** and monitor students' work, you may want to ask questions that prompt students to explain why their answer makes sense. Allow time for student pairs to share their thinking. Encourage students to use a variety of methods such as reasoning, using the mini-clocks, making pictures, or writing numbers to prove their answer makes sense.

Figure 13.4 Which One Doesn't Belong?

A	B
three-thirty	
C	D
3:30	**half past three**

5. Bring the class together for a whole-group discussion. Allow students to share their responses and reasoning for their choice. Record student responses. Encourage students to explain why one of the choices doesn't belong; students should notice that all four choices describe the same time.

6. Select student volunteers to show each time described in the prompt using the demonstration analog clock. Some students may have trouble accepting A or D as the same as the others. Ask, "What number is halfway around the clock from the hour? Where would the minute hand be placed to show it has traveled halfway around the clock from the hour?" to help facilitate student understanding of D. Also ask, "How many minutes are in one hour?"

7. Consider showing the time by creating a circular number line. Stretch a piece of string to show a traditional number line and ask, "Where is the halfway point?" Mark it with a clothespin or large paper clip and test it by folding the string in half. Then make a circle with the string and ask the students to notice where the halfway point is now. Then ask, "How many minutes are in a half of an hour?" Then ask, "What does it mean when we say that it is 3:30? Turn and tell your partner." Continue to change the representation of the number line by moving back and forth from the number line to the circular representation.

8. **Observe** as students discuss their thinking with their partners. Students' discussions might include understandings like the following: "Three-thirty is when the minute hand is on the 6 or halfway around the clock face and the hour hand is between 3 and 4."

FACILITATE

1. Read the task to students: "Zaara showed this time on her clock."

Figure 13.5 Zaara's Clock

Gabriel said that the clock shown in Figure 13.5 shows the time 10:30. Do you agree or disagree with Gabriel? Why? Show how you know."

2. Distribute the What Time Is It? student page. Allow time for students to discuss their thinking with a partner. **Observe** and monitor student discussions. Allow students to use the recording sheet to record their responses to the task question. Students may use words and/or pictures to show how they know.

CLOSE: MAKE THE MATH VISIBLE

1. Select student pairs to share their work. Students should have work that shows that they disagree with Gabriel. (The clock shows 9:30.) As students share their work, encourage them to use the language of "hour hand" and "minute hand" in their explanation. Have mini-clocks available for students to use to support their reasoning with a demonstration.

2. **Hinge Question.** "Whenever we set or look at an analog clock that represents time to the half hour, what can you say about the placement of the minute hand? The hour hand?" Students may need a reminder that the hour hand counts the hours that go by as it moves around the clock face.

3. **Extension.** "Write a letter to Gabriel explaining why the time on the clock isn't 10:30. Use what you know about the hour hand and minute hand to show him how to recognize the right time in your letter."

TASK 43: WHAT TIME IS IT STUDENT PAGE

 To download printable resources for this task, visit **resources.corwin.com/ ClassroomReadyMath/K-1**

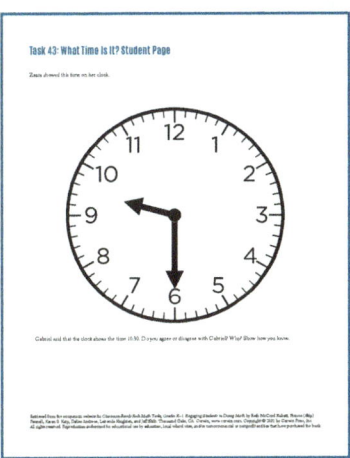

POST-TASK NOTES: REFLECTION & NEXT STEPS

Mathematics Standard

- Tell and write time in hours and half-hours using analog and digital clocks.

Mathematical Practices

- Use appropriate tools strategically.
- Attend to precision.

Vocabulary

- hand
- face
- digital
- analog
- display
- o'clock
- schedule

Materials

- student analog clocks with moveable hands, one per student
- pencil
- demonstration analog clock
- Fun Friday Schedule student page, one per student
- chart paper, marker

Task 44
Fun Friday Schedule

Write times using digital and analog clocks

TASK

Fun Friday

Figure 13.6 Jordan's List

Source: Maike Hildebrandt/iStock.com

Tomorrow is Fun Friday at Jordan's school. He made a list of things he plans to do (Figure 13.6) and the time each activity starts for his sister, Jasmine. He added clocks that show the digital and analog times to the schedule so that she won't be late. You and your partner will work together to help Jordan complete Jasmine's schedule by filling in the digital display and showing the same time on the clock face.

TASK PREPARATION

- Students will work in pairs for this task.

LAUNCH

1. Show students the digital time, 1:00. Say, "Can anyone tell me what time this display represents?" Allow time for student responses. Say, "You're correct. This display represents 1 o'clock."

ALTERNATE LEARNING ENVIRONMENT

To facilitate this task via distance learning, consider a digital clock such as the Math Clock app from the Math Learning Center: apps .mathlearningcenter.org/ math-clock/

2. Hold up an analog clock. Ask, "What will the clock face look like when it shows 1 o'clock? Show me this time on your clock." **Observe** as students arrange the hands to show 1 o'clock on their analog clocks. Make note of students who experience difficulty with this task.

3. Say, "Turn and talk to a partner. Tell your partner how you decided where to position the hour and minute hands to show 1:00." Allow time for students to share their responses.

4. Pose a few more sample times to students. Be sure to include times with half hours. Encourage students to explain how they decided where to place the hour and minute hands as they show each hour and half hour time.

ACCESS AND EQUITY

Providing students with opportunities to compare their clocks and discuss how they decided to organize the time allows students to develop a more accurate and complete understanding of time.

FACILITATE

1. Say, "Let's think about the activities we do during the day at school." List the activities named by the students on chart paper. Pose times of the school day that relate to the activities listed by students. Select one time that represents time to the hour or half hour. (If your school schedule doesn't include times to the hour or half hour, select a time to the hour or half hour that is closest to a time on your actual school schedule.) Write the time as a digital display. Say, "Show this time on your clock face."

2. **Observe** students as they work. Say, "We just showed the time we go to recess as a digital display and on an analog clock. Turn and talk to a partner and tell how you decided where to put the hour and minute hands." Select a few students to share their responses with the group.

3. Say, "Today, we're going to use what you know about showing times on an analog clock and a digital display to complete a schedule." Distribute copies of the task and clocks to students.

4. Read the task aloud as students follow along:

   "Tomorrow is Fun Friday at Jordan's school. He made a list of things he plans to do and the time each activity starts for his sister, Jasmine. He added clocks that show the digital and analog time to the schedule so that she won't be late. You and your partner will work together to help Jordan complete Jasmine's schedule by filling in the digital display and showing the same time on the clock face."

5. Group students in pairs. Allow time for students to work on the task. As students are working, visit each pair of students and **Interview** student pairs by asking questions such as: "How did you decide where to put the hour and minute hands to show each time? Why?" Make note of students who may have trouble with skip counting or confuse the hour and minute hands.

STRENGTHS SPOTTING

Allow students who demonstrate proficiency with placement of the hour and minute hands on this task to serve as peer coaches to share their strategies with their peers who may need additional support. Value multiple solution pathways and unique solutions, too.

6. Allow student pairs to join another pair of students for a *Pair-to-Pair Share*. Allow time for students to share their completed work, explain their answers, and ask questions if needed within their new groups.

CLOSE: MAKE THE MATH VISIBLE

1. Post completed schedules around the classroom.

2. Ask, "Were you correct? How can you prove you were correct? How did you decide where to put the hour and minute hands?" Record student responses.

3. **Exit Task.** Sasha is a first grader who is learning how to tell time. Explain to her how to show time on an analog clock. Make sure to tell her how you use the hour and minute hands to tell time. Show your thinking with words and pictures.

TASK 44: FUN FRIDAY SCHEDULE STUDENT PAGE

 To download printable resources for this task, visit **resources.corwin.com/ ClassroomReadyMath/K-1**

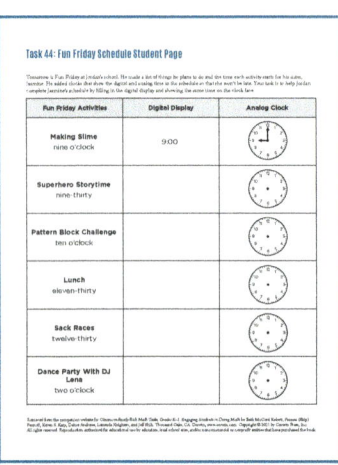

POST-TASK NOTES: REFLECTION & NEXT STEPS

_____

_____

_____

_____

_____

_____

Data

Represent and Interpret Data

TASK 45: GRADE 1: PIZZA PARTY

Organize, represent, and interpret data with up to three categories; ask and answer questions about the total number of data points, how many in each category, and how many more or less are in one category than in another.

TASK 46: GRADE 1: FIELD TRIP

Organize, represent, and interpret data with up to three categories; ask and answer questions about the total number of data points, how many in each category, and how many more or less are in one category than in another.

TASK 47: GRADE 1: SURVEY TIME

Organize, represent, and interpret data with up to three categories; ask and answer questions about the total number of data points, how many in each category, and how many more or less are in one category than in another.

Anticipating Student Thinking: Data are all around us. This chapter's tasks will engage first graders in data activities where they will organize, represent, and interpret data. The contexts of the chapter's tasks are votes for a class pizza party; determining the location for a class field trip; and creating, completing, and interpreting a survey about class favorites. These tasks will provide your students with their first opportunity to actually organize and interpret data using tables and related organizers as they apply number-related concepts previously learned (e.g., responding to "How many?" and "How many more?" questions). Your use of the classroom-based formative assessment techniques of Observation, Interview, and Show Me should be particularly helpful as you monitor student progress and understandings, given the introductory nature of organizing and interpreting data in a formal way at this grade level.

THINK ABOUT IT

The use of tables, charts, and other representations for collections of data is new for students at the first-grade level. Expect students to productively struggle as they gain experience in using such organizers for the data that they will interpret.

Mathematics Standard

- Organize, represent, and interpret data with up to three categories; ask and answer questions about the total number of data points, how many in each category, and how many more or less are in one category than in another.

Mathematical Practices

- Make sense of problems and persevere in solving them.
- Model with mathematics.

Vocabulary

- sort
- organize data
- label
- chart
- table
- most
- least
- compare
- categories
- data

Task 45
Pizza Party

Organize and interpret data

TASK

Pizza Party

Figure 14.1 Pizza

Source: guru86/iStock.com

Sasha asked her friends what kind of pizza (Figure 14.1) they wanted for the class pizza party. Their votes are in the table shown in Figure 14.2. Now that Sasha has all of their votes, she still isn't sure how much of each kind of pizza to order.

- Your task is to help Sasha organize the votes by making a chart to show how her friends voted.

- How can you use your chart to figure out how many students voted?

- Use the index cards to write two questions that can be answered by the data. Write each question on the front of a card.

- Then answer each question on the back of the card using the data in this chart and show how you determined your answer.

Figure 14.2 Pizza Chart

Pepperoni	Cheese	Veggie	Cheese	Pepperoni
Cheese	Pepperoni	Pepperoni	Cheese	Pepperoni
Veggie	Cheese	Veggie	Pepperoni	Cheese
Pepperoni	Pepperoni	Cheese	Veggie	Cheese
Cheese	Veggie	Pepperoni	Pepperoni	Veggie

ALTERNATE LEARNING ENVIRONMENT

To provide access to students in remote learning settings, use Google Classroom to assign a Google doc such as this example for student access to the task: Favorite Type of Cookie poster, https://docs.google.com/presentation/d/16RMrGwTyx6pDp42cVD5AnNjbRMP7zMt6XTH3ecYE-zw/edit?usp=sharing

TASK PREPARATION

- Students should have plenty of informal experiences with interpreting, sorting, and organizing data prior to completing this task.

- Prepare the poster for the Launch activity ahead of time. You'll need chart paper or a poster and sticky notes.

- Students will need access to markers and/or crayons to create their charts for the Pizza Party task.

- It would be helpful to prepare sets of markers or crayons for each pair of students in advance.

ALTERNATE LEARNING ENVIRONMENT

To provide access to students in remote learning settings, use Google Classroom to assign a Google doc such as this example for student access to the task. The linked slide allows students to reorganize the pizza types in order to better interpret the data. Pizza Types Cards, https://docs.google.com/presentation/d/1fVv6IdMrf9UWuDFHTk4pb_PUCgMa_Mg6JjHZSJbE32A/edit?usp=sharing

Materials

- prepare a large copy of the data from the Pizza Party task or make plans to display a copy using a document camera

- large (12" × 18") piece of construction paper, one piece per pair of students

- markers and/or crayons, one set per pair of students

- Pizza Party student page, one per pair of students

- index cards, two per pair of students

- sticky notes for Launch activity

- provide students with chart paper for the partner activity and have access to sticky notes in case students want to use this method to display their data (optional)

- Pizza Types Cards student page (optional)

LAUNCH

1. Bring students to a whole-group discussion. Say, "I asked some of my friends to vote for their favorite type of cookie: chocolate chip, oatmeal, or sugar. This poster shows the results." Prepare a similar poster using sticky notes labeled with the types of cookies (chocolate chip, 5; oatmeal, 3; and sugar, 4) and show it to students (see Figure 14.3). Facilitate a See, Think, and Wonder with students. Record student responses.

Figure 14.3 Favorite Types of Cookies

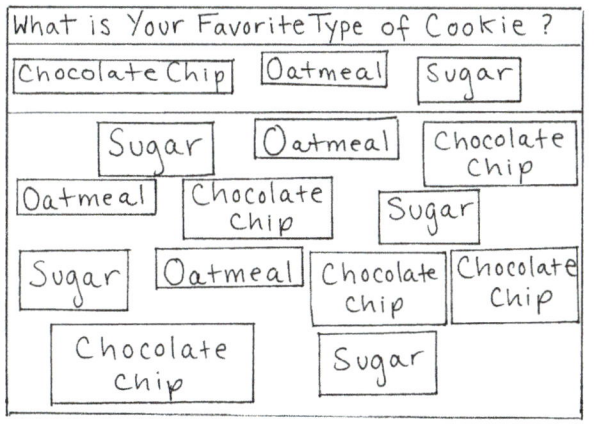

2. Say, "I want to compare the categories and find out which type of cookie most of my friends selected as their favorite. How could we organize the sticky notes so that we can count and compare the categories easily? Turn and talk to a partner."

3. Allow time to record and discuss student responses. Students may suggest sorting the sticky notes then lining up the sticky notes so they can compare the amount in each category. Select a few student volunteers to organize the sticky notes to resemble a bar graph. It may be necessary to remind students to organize the sticky notes neatly without any gaps or overlaps to make it easier to compare the categories. Place a line for students to place the sticky notes and build the graph.

4. Then, ask the following:

   » How did lining up the sticky notes help you compare the categories?

   » What do you know about my friends' favorite cookie type that you didn't know before we lined up the sticky notes? (Allow time for students to share their thinking. Record student responses.)

   » How did organizing the information on the poster help you think about this information? (Record student responses.)

   » What is a question we could ask and answer based on the information on the poster? (Record the student-generated questions.)

FACILITATE

1. Say, "We used sticky notes to organize the data about a favorite type of cookie. Can anyone think of a different way we could have organized the data?" Facilitate a *Think-Pair-Share*. Record student responses. (Student responses may include using numbers in a table, drawing pictures, making an X for each response, or making tally marks.) Say, "Now, you and a partner will sort, organize, and answer questions about a new set of data."

ACCESS AND EQUITY

Students benefit greatly from discussions with their peers. Allowing students individual think time allows each student to communicate their own ideas about organizing data. Sharing ideas with a peer and then with the group provides students with access to multiple strategies for organizing data before engaging in the task.

2. Introduce the task. Say, "Sasha asked her friends what kind of pizza they wanted for the class pizza party. Their votes are in the table below. Now that Sasha has all of their votes, she still isn't sure how much of each kind of pizza to order."

3. Organize the students into pairs and distribute materials.

4. Say, "Your task is to help Sasha organize the votes by making a chart to show how the friends voted. What questions could we ask and answer based on the way you organized the data?" **Observe** students as they work. Circulate among the student pairs asking the following **Interview** questions:

   » How did you organize the data?

   » How did you and your partner arrange the sticky notes to make it easier to compare the categories?

   » How can you use your chart to figure out how many students voted? How is making a chart helpful when we're looking at a lot of information?

   » Why is sorting and organizing data important?"

5. Say, "Use the index cards to write questions that can be answered by the data. Each partner writes a question on the front of a card. Then trade cards and answer each question on the back of the card and show how you got the answer. Trade cards to check each other's work."

6. Allow each pair of students to join another pair for a *Pair-to-Pair Share*. Student pairs will take turns presenting their questions to each other and challenging their classmates to answer their questions.

7. Then, ask the students:

   » How did organizing the pizza types help you answer questions about Sasha's friends?

   » Can you use your chart to answer your new group members' questions?

8. Provide time for students to share their completed work, ask and answer questions, and explain their responses.

9. **Observe** the pairs as they share. Record the methods the student pairs used to organize and display the data. Look for the following:

   » Did students arrange the pizza types in rows or columns side by side to make comparing easier?

   » Did students write questions that can be answered using the data?

   » What type of solution strategies did students use to answer the questions?

   » Student pairs who used different methods such as lining up objects, tally marks, numbers in a table, or pictures.

10. Ask student pairs who used different methods to prepare to share in whole group.

Note: Consider using the Observation and Interview tools for recording student responses (see Appendix B).

CLOSE: MAKE THE MATH VISIBLE

1. Bring students to a whole-group discussion. Facilitate a Something Similar and Something Different Gallery Walk. Allow the selected student pairs to share their displays of the data. Allow time for students to discuss similarities and differences in the data displays. Ask questions that prompt students to compare the advantages and disadvantages of using different ways to display the data: "How could you use pictures, sticky notes, tally marks, or numbers? What can we see with pictures/drawings that we can't see with numbers or tally marks?"

2. **Hinge Question.** Ask, "In what ways do graphs help us see information in a quicker and easier way?"

3. Discuss the questions from the pizza party task and allow students to share their responses and supporting work. Ask, "Did organizing the data help you find out what kind of pizza Sasha should buy the most of? The least of? How did organizing the data help you figure out how many friends voted? Show how." Allow time for students to respond. Record any additional information and add to the list from the Launch activity. Collect the index cards and select student pairs to ask their questions to the group. Allow time for students to both discuss and show how they answered the questions. Observe how they use the data collected to answer the questions.

TASK 45: PIZZA PARTY STUDENT PAGES

 To download printable resources for this task, visit **resources.corwin.com/ ClassroomReadyMath/K-1**

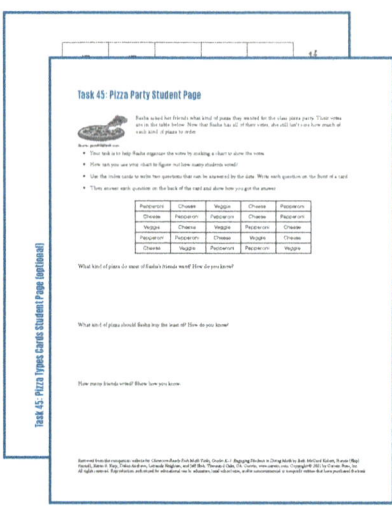

POST-TASK NOTES: REFLECTION & NEXT STEPS

Task 46
Field Trip

Interpreting data

TASK

Field Trip

Mrs. King's first-grade class voted on their favorite place to visit for a field trip. Each student voted once. Figure 14.3 shows the results.

Figure 14.3 Favorite Field Trip

Field Trip	Number of Students
Aquarium	11
Children's Museum	6
Zoo	8

The students wrote their own questions.

- Malia asked, "How many more people chose the Aquarium than those who chose the Zoo?"

- Justin asked, "How many more people chose the Aquarium than those who chose the Zoo and the Children's Museum combined?"

- Andrea said that you need to look at all three field trip choices to answer both questions. Do you agree or disagree with Andrea? Explain your answer.

TASK PREPARATION

- Copy and cut apart the Question Cards and Favorite Sport Chart for the Launch activity.

- Each pair of students will need one Question Card and one copy of the Favorite Sport Chart.

- Students will work in pairs for this task. Decide how to pair students prior to the task.

LAUNCH

1. Bring students to whole-group discussion. Show students the chart that follows (Figure 4.4). Facilitate a Notice and Wonder about the data set and allow students to share their thinking.

Mathematics Standard

- Organize, represent, and interpret data with up to three categories; ask and answer questions about the total number of data points, how many in each category, and how many more or less are in one category than in another.

Mathematical Practices

- Construct viable arguments and critique the reasoning of others.
- Attend to precision.

Vocabulary

- data
- equation
- compare

Materials

- Favorite Sport Chart student page, one per pair of students
- Question Cards for Launch student page, one card per pair of students
- Field Trip Recording Sheet student page
- blank paper for showing work, several pages per pair
- pencils, one per student
- counters or connecting cubes (make available to students for comparing tasks)

Favorite Sport

A group of students voted for their favorite sport to play at recess. Figure 14.4 shows their votes.

Figure 14.4 Recess Votes

Sport	Number of Votes
Baseball	7
Basketball	12
Soccer	11

2. Ask, "Can you think of a question you could ask about this information? Turn and talk to a partner to discuss your thinking." Have students share their thinking.

3. Organize the students into pairs. Say, "Today we're going to practice answering questions based on data. Let's try one together. How could we figure out how many students prefer baseball and soccer combined? What information would you use from the data table? Turn and talk to your partner." Provide time for the students to discuss their thinking.

4. Next ask, "Can you write an equation to find the answer? Show me how you found the answer." Encourage student pairs to share their strategies. Students should share that they solved the problem by adding 7 + 11 to determine that 18 students prefer baseball *and* soccer.

5. Say, "You and your partner will receive one question to answer using these data." Distribute one Question Card to each pair of students. (It's okay if more than one pair has the same question.) Tell students, "Make sure you show your work and explain how you found the answer to your question." **Observe** students as they answer their assigned questions. Make note of how students organize the data to help answer the questions. Do they make a different representation or reorganize the data (pictures, drawings, tally marks, manipulatives, or bar graphs) to help make sense of the data?

6. **Interview.** Meet with each pair and ask questions about what information the pairs need to answer their question and their plan to answer the question.

Note: Consider using the Observation and Interview tools to record student responses (see Appendix B).

> **PRODUCTIVE STRUGGLE**
>
> Some of the Question Cards may be more challenging than others. Select and assign questions as needed, keeping in mind which questions provide the appropriate level of struggle for each student.

7. Use *Stand Up, Hand Up, Pair Up* and allow student pairs to stand up, raise their hands, and pair up with another pair of students. Allow the student pairs to share their questions and solutions with each other.

FACILITATE

1. Bring students together as a group. Allow time for students to share their analysis of the sports data, including the types of questions they answered and the strategies they used. Record student responses on chart paper. If space allows, post student solutions from the Launch activity. (Students can use these as a reference.)

2. Say, "We just answered questions about data. I want to share information about some students who are writing their own questions about data." Divide students in pairs. Introduce the task to students: "Mrs. King's first-grade class voted on their favorite place to visit for a field trip. Each student voted once. This table shows the results." (Project a copy of the field trip chart for students to see.) "Turn and talk to your partner and share one question you could ask about these data." Select students to share their questions.

3. Distribute recording sheets, pencils, and any other materials students may need to support their work. Share the rest of the task with students: "The students wrote their own questions."

   » Malia asked, "How many more people chose the Aquarium than those who chose the Zoo?"

   » Justin asked, "How many more people chose the Aquarium than those who chose the Zoo and the Children's Museum combined?"

   » Andrea said that you need to look at all three field trip choices to answer both questions. Do you agree or disagree with Andrea? Explain your answer.

4. **Observe** as students work. Remind students: "Read each question. What information do you need to answer Malia's and Justin's questions? How would you solve each question? How could you use what you know about answering each question to help decide if you agree or disagree with Andrea? Use this information to find out if you agree or disagree with Andrea. Explain your answer and show how you know."

5. Once student pairs have answered the task question, students will participate in a modified *One Stay, The Other Stray*. One partner will stay to explain the pair's work, and the other partner will stray to another partner to find out how they solved the task. Allow time for students to share their completed work, explain their answers, and ask questions. **Observe** pairs as they share.

6. Students return to their partner and share.

Note: Consider using the Observation tool for recording pair responses (see Appendix B).

ACCESS AND EQUITY

Students demonstrate understanding of math concepts when they're allowed to make connections and apply them in a variety of situations. This task allows students to use what they know about solving addition and subtraction problems while interpreting data.

ACCESS AND EQUITY

Students learn from engaging in multiple opportunities for productive discourse. Using the modified One Stay, The Other Stray allows students to talk to a peer and discuss their mathematical understanding. Speaking to a partner allows students to problem solve and discuss their thinking before sharing with the whole group.

CLOSE: MAKE THE MATH VISIBLE

1. Bring students to circle time. Display the field trip chart. Reread the task and pose the question to students: "Do you agree or disagree with Andrea?" Select student pairs to share and explain their work. Ask, "How did you use the data to answer the task question? Did anyone write an equation to help answer the question?" Allow time for students to share.

2. **Exit Task.** Have students examine the data in Figure 14.5 and write and answer two compare questions about the data.

Figure 14.5 Favorite Movie Theater Snack

Snack	Votes
Popcorn	15
Candy	13
Hot Dog	7

TASK 46: FIELD TRIP STUDENT PAGES

online resources — To download printable resources for this task, visit **resources.corwin.com/ ClassroomReadyMath/K-1**

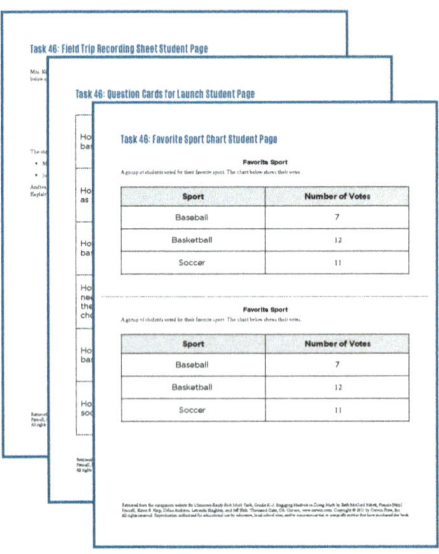

POST-TASK NOTES: REFLECTION & NEXT STEPS

Task 47
Survey Time

Represent and interpret data

TASK

Survey Time

It's survey time—you will ask your classmates to answer a question that you want to know more about!

1. Choose a question from the list provided on the Survey Time student page. Color the box for the question you chose.

2. Choose three answer choices to include on your survey. Write the answer choices on the recording sheet.

3. Ask your classmates the question and record their answers on the recording sheet.

4. Collect and organize the data.

5. Make a poster.

6. Answer questions about your data. Explain how you know.

TASK PREPARATION

- Students should have multiple opportunities with sorting and organizing data as a prerequisite to this task.

- Assemble all necessary materials.

- Make sure that students can easily move around the room to gather responses to their questions.

- Prepare a procedure for gathering the survey responses so that students collect data from classmates in an orderly manner. One suggestion is to allow student groups to "partner" with another group and take turns posing their questions to each other. Once each group has posed their questions and collected the responses, keep repeating the process until each group has collected data from all classmates or at least half the class.

LAUNCH

1. Bring students to a whole-group discussion. Tell a story of your own making, or, if the book is available, read *The Best Vacation Ever* by Stuart J. Murphy. Discuss the story details with students. In this story from the book

Mathematics Standard

- Organize, represent, and interpret data with up to three categories; ask and answer questions about the total number of data points, how many in each category, and how many more or less are in one category than in another.

Mathematical Practices

- Reason abstractly and quantitatively.

- Model with mathematics.

Vocabulary

- data

- compare

- category

- sort

Materials

- book, *The Best Vacation Ever* by Stuart J. Murphy

- chart paper, one piece per group of three

- markers or crayons, one set per group of three

- Recording Sheet student page, one per group of three

- Survey Questions student page

- sticky notes, one set per group of three

- counters or connecting cubes (provide access to all students)

- tape to display the posters

the family surveyed all family members to decide where to go on vacation. Sample questions to ask students include the following: "What did the family do to decide where to go on vacation? How did they sort and organize their data? How did they use the sorted data to make a decision?"

Note: The following Launch activities may be completed with or without the children's literature experience noted above.

2. Say, "Think of a time when you used a survey or asked a question to get information." Allow time for students to share. If students have experience in this area, allow them to share briefly. Students might share a time when the family decided where to go to dinner and family members shared their preferences.

3. Tell students, "Today you are going to complete a survey to find out the answer to a question. Think of a question you would like our class to answer." Record student responses (Figure 14.6). Say, "I have a question I want to ask the class. What is your favorite season? The choices are spring, summer, fall, and winter." Display the question on chart paper and model how to quickly collect the information (e.g., use tally marks, write the numbers in a table, make Xs, draw circles, use sticky notes).

Figure 14.6 Tallying Student Responses

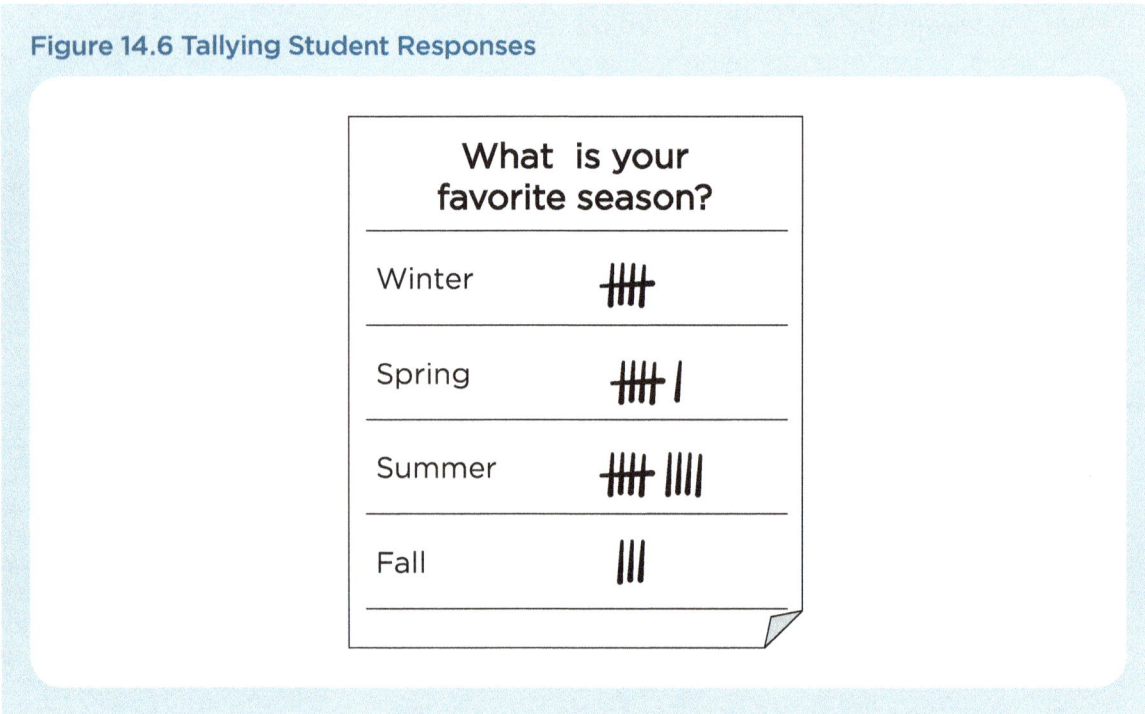

4. Say, "Turn and talk to a partner and describe how you could organize and show this information on a poster." Record student suggestions on chart paper.

5. Ask, "What can we learn looking at this information? Let's make a list of other ways we can organize and show the data on a poster. Can you think of some of the other ways we've organized data?" Add student suggestions to the poster.

FACILITATE

1. Say, "You helped me quickly collect information about the class to answer my question. Now you are going to ask your classmates a question, collect the data, and organize the data using a poster. You can use one of the ways we used on our class poster or think of your own way to show the data. Then you'll have the opportunity to answer some questions."

2. Divide students into groups of three. Distribute the Survey Time recording sheet. Read the directions with students:

It's survey time—you will ask your classmates to answer a question that you want to know more about!

   1. Choose a question from the list below. Color the box for the question you chose.

   2. Choose three answer choices to include on your survey. Write the answer choices on the recording sheet.

   3. Ask your classmates the question and record their answers on the recording sheet.

   4. Collect and organize the data.

   5. Discuss what the data are saying.

   6. Make a poster.

   7. Answer questions about your data. Show how you know.

3. **Observe** the student groups as they collect, sort, and organize their data. Make notes of procedures students use to collect the data. For example: Do students use numbers, pictures/drawings, or tally marks? What methods do they use to compare categories in order to answer questions? Can students easily organize the data?

Note: Consider using the Observation tool for recording the responses (see Appendix B).

4. **Interview.** As you observe the student groups, consider using one or more of the following interview questions:

   » How did you organize the data?

   » How can you organize the information to see your data more clearly?

   » How many students chose each category?

   » How could you have used tally marks or sticky notes? How did using sticky notes, cubes, or tally marks help you answer the questions?

   » What did you find out?

   » What method(s) did you use to answer the compare questions?

   » Write an equation to show how you found your answer to the compare questions. How did you solve question 3? How did you solve question 4? How did you solve question 5?

5. Have students engage in a Group Rotate and Share. Each group rotates to another group, and both groups share their questions, data displays, and the answers to their questions. Allow time for students to share their results, ask questions, and share what they noticed. Tell students, "After meeting with your new group members, make any changes or add any new information to your poster."

PRODUCTIVE STRUGGLE

It's important to resist the temptation to step in too soon to provide guidance on an open-ended task. If students need additional scaffolds, remind them of the chart generated within the task's Launch. Or suggest that students use sticky notes or connecting cubes to represent student choices to make it easier to compare quantities.

ACCESS AND EQUITY

Student choice allows students to play an active role and choose the materials and methods that best meet their needs. Some students will choose more complicated methods than others, but this task allows all students to be successful using the materials and methods that best meet their needs.

CLOSE: MAKE THE MATH VISIBLE

1. Post the student data representations around the classroom. Allow students to complete a Gallery Walk to view all of the completed posters. Give each student a sticky note and allow students to write a question or comment to one group's poster.

2. Bring students to whole-group discussion. Allow students to discuss and debrief the experience as a class. Ask, "How did you organize the data? How can you organize information to see data more clearly? How many of your groups used tally marks? Sticky notes? Numbers? Cubes?" (Allow students to indicate their method using a show of hands.) "Did one method make it easier or harder to answer questions about the data? What method did you use to solve the compare problems?" Select a few student groups to show their work for problems 3, 4, and 5. Discuss similarities and differences in solution methods.

3. As a culminating activity, allow each group to select a question to ask another grade level. Students can then visit another grade-level class, ask their questions, and collect and organize the data.

TASK 47: SURVEY TIME STUDENT PAGES

online resources — To download printable resources for this task, visit **resources.corwin.com/ ClassroomReadyMath/K-1**

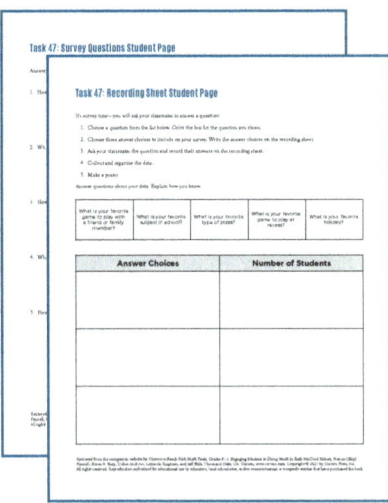

POST-TASK NOTES: REFLECTION & NEXT STEPS

_____

_____

_____

Geometry

Distinguishing Attributes and Naming and
Identifying Shapes

TASK 48: KINDERGARTEN: I SPY

Describe objects in the environment using names of shapes, and describe the relative positions of these objects using terms such as *above, below, beside, in front of, behind,* and *next to.*

Correctly name shapes regardless of their orientations or overall size.

TASK 49: GRADE 1: ALWAYS OR SOMETIMES TRUE

Distinguish between defining attributes (e.g., triangles are closed and three-sided) versus non-defining attributes (e.g., color, orientation, overall size); build and draw shapes to possess defining attributes.

TASK 50: GRADE 1: WHAT IT IS AND WHAT IT'S NOT

Distinguish between defining attributes (e.g., triangles are closed and three-sided) versus non-defining attributes (e.g., color, orientation, overall size); build and draw shapes to possess defining attributes.

TASK 51: KINDERGARTEN: THE *TILTED* TRIANGLE

Correctly name shapes regardless of their orientations or overall size.

TASK 52: KINDERGARTEN: FLAT OR NOT FLAT?

Identify shapes as two-dimensional (lying in a plane, "flat") or three-dimensional ("solid").

Anticipating Student Thinking: Preschool students have many informal experiences with geometry. They play games involving two- and three-dimensional shapes, and they literally see and experience attributes related to shapes before they enter kindergarten. Tasks at the kindergarten and first-grade levels engage students in activities designed to help develop spatial sense. The task lessons in this chapter focus on spatial relationships and emphasize position (e.g., above, below, behind, beside, etc.); attributes of shapes; determining shapes regardless of their orientation; and differences between two- and three-dimensional shapes. Note that the shapes around your classroom and in the homes of your students (e.g., boxes, blocks, balls, cones) can be used informally. Such opportunities will help to develop student ability to visualize and manipulate objects in space.

THINK ABOUT IT

Ensure that all of your students have opportunities to engage with shapes and have the opportunity to discuss the attributes of shapes. Such discussions will strengthen the reasoning of your students and may range from determining which shapes are bigger or smaller than others, to specific differences between two- and three-dimensional shapes.

Mathematics Standards

- Describe objects in the environment using names of shapes, and describe the relative positions of these objects using terms such as *above, below, beside, in front of, behind,* and *next to.*

- Correctly name shapes regardless of their orientations or overall size.

Mathematical Practice

- Attend to precision.

Vocabulary

- spy
- circle
- triangle
- square
- oval
- rectangle
- hexagon
- cube
- cone
- cylinder
- sphere
- above
- below
- behind
- in front of
- next to
- beside

Materials

- Assorted Shapes to Find student page
- manipulatives (optional)

Task 48
I Spy

Describe the position of shapes in the environment

TASK

I Spy

Figure 15.1 Looking for Shapes

Source: Olga Kurbatova/iStock.com

Yesterday after school, [insert name of school mascot or have the class identify a school mascot] came into our classroom and made a game for us! [The mascot] hid a bunch of different shapes around the classroom. Our job is to find all of the shapes (see Figure 15.1).

TASK PREPARATION

- Cut out shapes for the activity. If you are unable to print using a color copier, consider quickly shading in the figures with assorted colors before cutting them out. This will help students describe each shape they find (e.g., "I found the *green* triangle behind the bookshelf").

- Consider using physical manipulatives instead of the images for three-dimensional shapes.

- "Hide" the shapes in various locations around the classroom.

LAUNCH

1. Bring the class together and present the task. Facilitate a See, Think, and Wonder, and record students' ideas.

2. Ask, "What kinds of shapes do you think [the mascot] might have hidden around the room? What are some places you think shapes might be hidden?"

Figure 10.2 Recording Students' Ideas

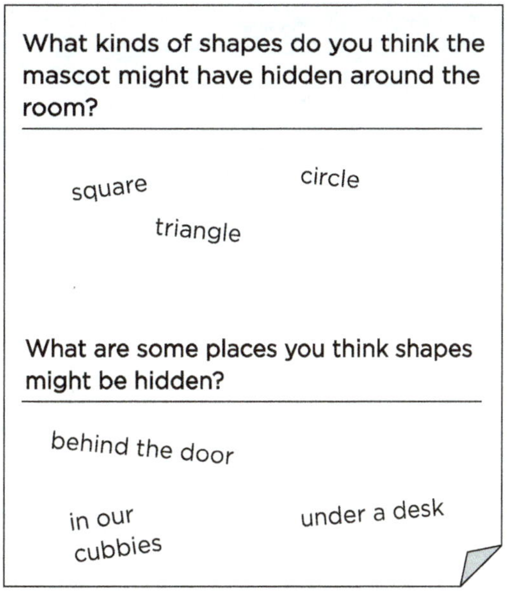

> **What kinds of shapes do you think the mascot might have hidden around the room?**
>
> square   circle
>   triangle
>
>
> **What are some places you think shapes might be hidden?**
>
> behind the door
>
> in our   under a desk
> cubbies

3. Record students' ideas on the board (see Figure 15.2).

4. Ask students to *Turn and Learn* from their partner how they will know if they found a triangle. What will it look like?

5. Ask students, "I might spy a shape that is *behind* something. What are some other words I could use to describe where a shape might be located?"

6. Elicit words and phrases students might use to describe relative position or location. Write these on the board as students share them.

FACILITATE

1. Organize students into heterogeneous pairs.

2. Give the student pairs time to search around the room to locate the hidden shapes. Remind them that they should not move the shapes they find or tell others that they found one.

3. As students circulate, conduct **Interviews** with student pairs. Ask,

   » Did you spy any triangles? Where?

   » How can you tell me without showing me where the shape is located?

   » Use one of the position words we thought about to describe where that triangle is located?

Note: Consider using the Interview (small group) tool for organizing and recording the pair responses (see Appendix B).

CLOSE: MAKE THE MATH VISIBLE

1. After you've had an opportunity to interview each pair, and most pairs have found several shapes, bring the class back together.

2. Partner each pair up with another pair and have students *Pair-to-Pair Share* about the shapes they found.

3. Ask, "Did you spy some of the same shapes? What are some of the shapes we spied?"

4. Allow each quartet to share a shape that was found. As they share, ask questions to elicit positional language, such as "Where did you find that shape? Tell me where I should look."

STRENGTHS SPOTTING

Look for students' communication strengths as they describe multiple attempts to spy shapes. Strengths spotting should focus on the actual disposition or content rather than right answers. Help students see what their strengths are by calling explicit attention to the strength they are exhibiting.

5. When each shape is located, ask the whole class to repeat the words the quartet used to describe its location. For example, "Turn and tell your partner! They found the red triangle *next to* the bookshelf!"

TASK 48: I SPY STUDENT PAGE

 To download printable resources for this task, visit **resources.corwin.com/ ClassroomReadyMath/K-1**

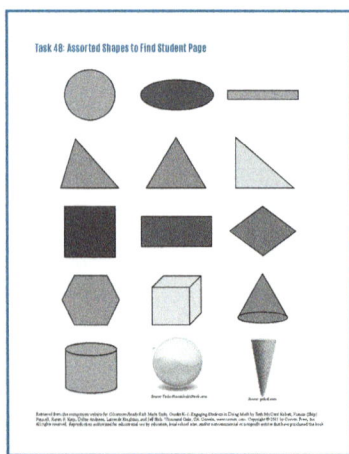

Mathematics Standard

- Distinguish between defining attributes (e.g., triangles are closed and three-sided) versus non-defining attributes (e.g., color, orientation, overall size); build and draw shapes to possess defining attributes.

Mathematical Practices

- Construct viable arguments and critique the reasoning of others.
- Look for and make use of structure.

Vocabulary

- geometry
- geometric
- shape
- always
- sometimes
- triangle
- square
- rectangle
- circle
- octagon
- cone
- cylinder
- cube
- attribute
- defining attribute
- non-defining attribute
- two-dimensional
- three-dimensional

Task 49

Always or Sometimes True

Understand the difference between defining and non-defining attributes

TASK

Always or Sometimes True

Mrs. Evans's first graders are studying attributes of shapes. Lucas said, "I think some attributes of shapes are *always* true, but other attributes of shapes are *sometimes* true."

Mrs. Evans has asked for our help to investigate! Each of your groups will analyze a different shape. Your job is to find which attributes are *always* true about your shape and which attributes are only *sometimes* true about your shape.

TASK PREPARATION

- Organize students into heterogeneous groups of two or three.
- Gather a collection of writing utensils and create a physical model like the one in the task image. Make sure all the pencils in the container are yellow and have erasers
- Gather a collection of pencils that are not yellow and/or do not have erasers.
- Prepare sorting collections, or create collections of physical objects for sorting.
- Prepare Attribute Cards. Be sure that no group gets a card that is *never true* for their shape. For example, be sure the card "have curves" is not given to any of the groups with polygons.

LAUNCH

1. Facilitate a Notice and Wonder using the image shown in Figure 15.3 or a real container of writing utensils.

Figure 15.3 Container of Utensils

Source: fcafotodigital/iStock.com

Materials

- writing utensil collection

- Shape Collection student pages (triangles, rectangles, squares, octagons, cylinders, circles, cones, and cubes), one set per group

- Attribute Cards student pages, one set per group

- Attribute Work Mat student page, one per group

- glue sticks or tape

- sticky notes, for the Gallery Walk

2. Ask, "What are some things you see in the container? How do you know that is a (pen/pencil/marker/etc.)?"

3. Record students' ideas on the board.

4. Tell students the following story:

   Mrs. Evans asked the class to count the number of pencils in the mug on her desk. The students said, "That's easy! There are four pencils in the mug!" Mrs. Evans said, "Wow, that was quick! How did you figure that out?" They said, "It was easy. We just looked for everything that is yellow and has an eraser."

5. Make connections to students' ideas about what was in the container. Ask, "Did you notice that the pencils were all yellow? What else did you notice (they have erasers)? Do you agree with what the class said? Are pencils always yellow, or only sometimes yellow?"

6. Ask, "Are pencils *always* yellow? Do pencils *always* have erasers?"

7. If students have counterexamples readily accessible, allow them to prove their thinking by showing the counterexamples to the class. Otherwise, show students a prepared collection of pencils that are not yellow and/or do not have erasers.

8. **Hinge Question.** How do we know these are pencils and not crayons, for example?

9. Display a copy of the Attribute Work Mat labeled "pencil attributes" and elicit student ideas for things that are "always" or "sometimes" true about pencils. Figure 15.4 shows what this attribute work mat might look like.

Figure 15.4 Attribute Work Mat for Pencils

<u>pencil</u> attributes

Always	Sometimes
has lead is used for writing	is sharp is yellow has an eraser

10. Present the initial part of the task prompt to students:

> Mrs. Evans's first graders are studying attributes of shapes. Lucas said, "I think some attributes of shapes are *always* true, but other attributes of shapes are *sometimes* true." Mrs. Evans has asked for our help to investigate!

FACILITATE

1. Organize students into pairs or small groups and distribute a different shape collection to each group.

2. Ask students to look at their collection of geometric shapes and say what they have (e.g., triangles, squares, etc.).

3. Students should write the name of their shape at the top of their Attribute Work Mat.

4. Give each group a set of Attribute Cards.

5. Model how they will take turns drawing an Attribute Card, say whether that attribute is always true or only sometimes true for that shape, and place the Attribute Card in the right spot on the Attribute Work Mat. For example, a student in the group with rectangles might draw the Attribute Card "has all sides the same length." The group would analyze their collection and then the student might say, "Rectangles *sometimes* have all sides the same length" and place the card in the "Sometimes" part of the mat. Then the next student might draw the Attribute Card "are two-dimensional" and say, "Rectangles are *always* two-dimensional" and place the card in the "Always" part of the mat as a defining attribute. Figure 15.5 shows what this attribute work mat for rectangles might look like.

ACCESS AND EQUITY

Encourage students to ask for any attribute to be read aloud as needed. The goal of this activity is reasoning about the defining attributes of the figures, not learning to read the vocabulary involved. Be available to read the attributes aloud at student request, but do not make preemptive decisions to read them all for any student. Allowing students to have control of that process will ensure we don't limit students' opportunities by reading cards for them that they might have confidently read on their own.

Figure 15.5 Attribute Work Mat for Rectangles

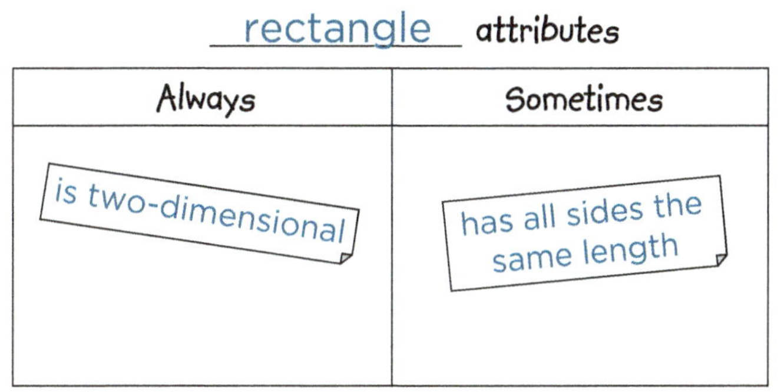

rectangle attributes

Always	Sometimes
is two-dimensional	has all sides the same length

6. **Observe.** As you circulate from group to group, take note of students' reasoning. As needed, prompt students to show evidence from their collections to support their placement of each Attribute Card.

7. **Show Me.** Ask each group specific questions about the cards they've already placed. Encourage students to defend cards their partners placed rather than only ones they placed themselves. For example,

   » Show me how you could sort your rectangles into rectangles with sides that have the same length and rectangles that don't have sides the same length.

   » Show me how you know that circles are _always_ two-dimensional. Show me what something that is _not_ two-dimensional looks like.

8. Once a group has a completed sort and all students in the group agree on the placement of the attributes, direct them to affix (glue/tape) the cards to the Attribute Work Mat. Place completed sorts on display. Early finishers can begin a new sort in a different space.

Note: Consider using the Observation and Show Me tools for recording student responses (see Appendix B).

> **! PRODUCTIVE STRUGGLE**
>
> Be careful to ask students to prove their thinking when a card is correctly placed and when a card is incorrectly placed. If we only ask students to defend their thinking when they are incorrect, we unintentionally encourage students to depend upon the teacher as the sense maker and often miss opportunities to provoke thinking where students may have found the correct answer using incorrect reasoning.

CLOSE: MAKE THE MATH VISIBLE

1. When all groups have completed at least one sort, conduct a Notice and Wonder Gallery Walk. Display Attribute Work Mats for each shape in the same space (e.g., display all triangle sorting mats together).

2. Review students' sticky notes and discuss themes with the class. Highlight the following:

   » Some of the shapes we sorted today were two-dimensional and some were three-dimensional.

   » Some attributes can be _sometimes true_ about one shape but are _always true_ about another shape. For example, rectangles _sometimes_ have all sides the same length, but squares _always_ have all sides the same length.

» Every shape has some *Always* attributes. These are called defining attributes.

  » A figure has to have *all* defining attributes for a shape to be called by that shape name. For example, a figure can only be called a square if it is made of exactly four straight sides that are all the same length and it has four square corners.

  » A figure that doesn't fit any one of the defining attributes for a given shape can't be called by that shape name. For example, a figure that is described as not having exactly four sides cannot be a square (even if it has all the other defining attributes) because "has 4 sides" is a defining attribute of "square."

» Every shape also has *Sometimes* attributes. These are called non-defining attributes.

  » Different shapes can share non-defining attributes. For example, both squares and circles can be red.

  » Knowing a non-defining attribute of a figure is not enough to know how to name it. For example, knowing that a figure is small and blue will not help you know if the figure is or is not a triangle.

3. Take time to ensure that all defining attributes are listed for each shape that has been explored. Analyzing one set of shapes at a time, ask students to *Turn and Learn* from a neighbor. Ask, "Look at the list you made of attributes that are always true about (rectangles*). Do we have all the defining attributes represented here? Can we draw a shape that has all of these attributes but *is not* a (rectangle*)?"*name a shape

4. Have the class attempt to draw a figure that is not the given shape with all the defining attributes that have been identified. For example, if the rectangle group only had "has 4 sides" and "has 4 corners" in their *Always* list, the class will attempt to draw a figure with four sides and four corners that is not a rectangle.

STRENGTHS SPOTTING

This task is an opportunity for students to showcase their geometric and spatial strengths. As you observe the students, give explicit feedback to students about how they are risk taking and drawing their rectangles.

5. Using the class-generated shapes as examples, elicit from the class what additional attributes should be added to the *Always* category to complete the list of defining attributes for that shape.

6. Display students' posters around the room so they can be referred to as students study various geometric figures. New posters can be added as additional figures and their attributes and defining attributes are studied.

TASK 49: ALWAYS OR SOMETIMES TRUE STUDENT PAGES

To download printable resources for this task, visit **resources.corwin.com/ ClassroomReadyMath/K-1**

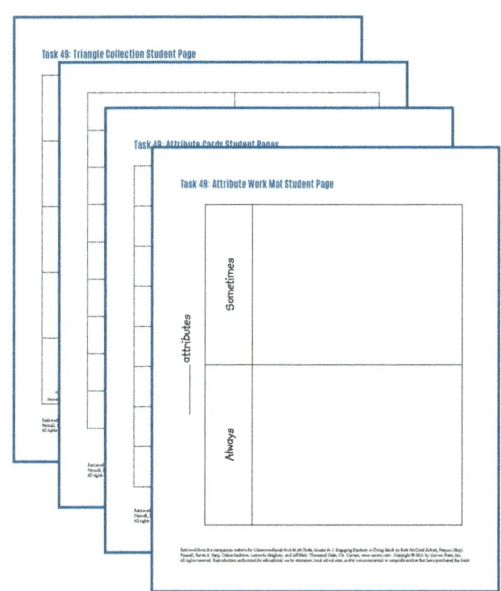

POST-TASK NOTES: REFLECTION & NEXT STEPS

Mathematics Standard

- Distinguish between defining attributes (e.g., triangles are closed and three-sided) versus non-defining attributes (e.g., color, orientation, overall size); build and draw shapes to possess defining attributes.

Mathematical Practice

- Construct viable arguments and critique the reasoning of others.

Vocabulary

- two-dimensional shapes
- defining attributes
- non-defining attributes
- rectangles
- squares
- trapezoids
- triangles

Materials

- Sorting Mat student page, at least one copy per student
- sticky notes (for closure)

Task 50

What It Is and What It's Not

Draw shapes that do and do not possess defining attributes

TASK

What It Is and What It's Not

Ora said, "I'm going to draw a square!" Damon said, "I'm going to draw a shape that is not a square!" Figure 15.6 shows what they drew.

Figure 15.6 Ora's and Damon's Shapes

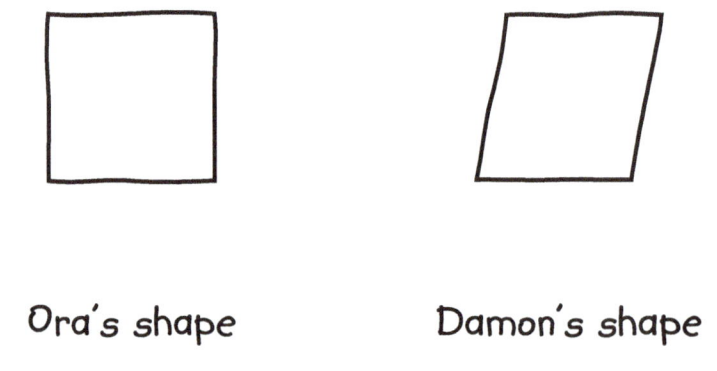

Ora's shape   Damon's shape

Explain why Ora's shape is a square but Damon's shape is not a square. Then work with a partner to draw your own What It Is and What It's Not pictures for other shapes!

TASK PREPARATION

- Use this task after students have learned about defining attributes of two-dimensional shapes.

- Decide which two-dimensional shapes you want students to use for the activity (e.g., rectangles, squares, trapezoids, triangles).

LAUNCH

1. Carefully sketch so that the drawing is accurate or project a variety of squares with varied size and orientation on the marker board.

2. Ask students to *Turn and Talk* with a partner. Say, "What do you know about the shapes I've shown on the board? How would you describe them? What name would you give these shapes?"

3. Bring the class back together and ask students to share their thinking.

4. **Hinge Question.** How do you know that these are all squares? (Elicit from the students that the shapes have four sides, etc.)

5. Present the task to the class.

6. Ask students to *Turn and Learn* from a partner. Remind students they should be prepared to share their partner's reasoning with the class.

7. Use reasoning provided by students to record students' ideas to complete a sample version of the Sorting Mat for "square like the one shown in Figure 15.7."

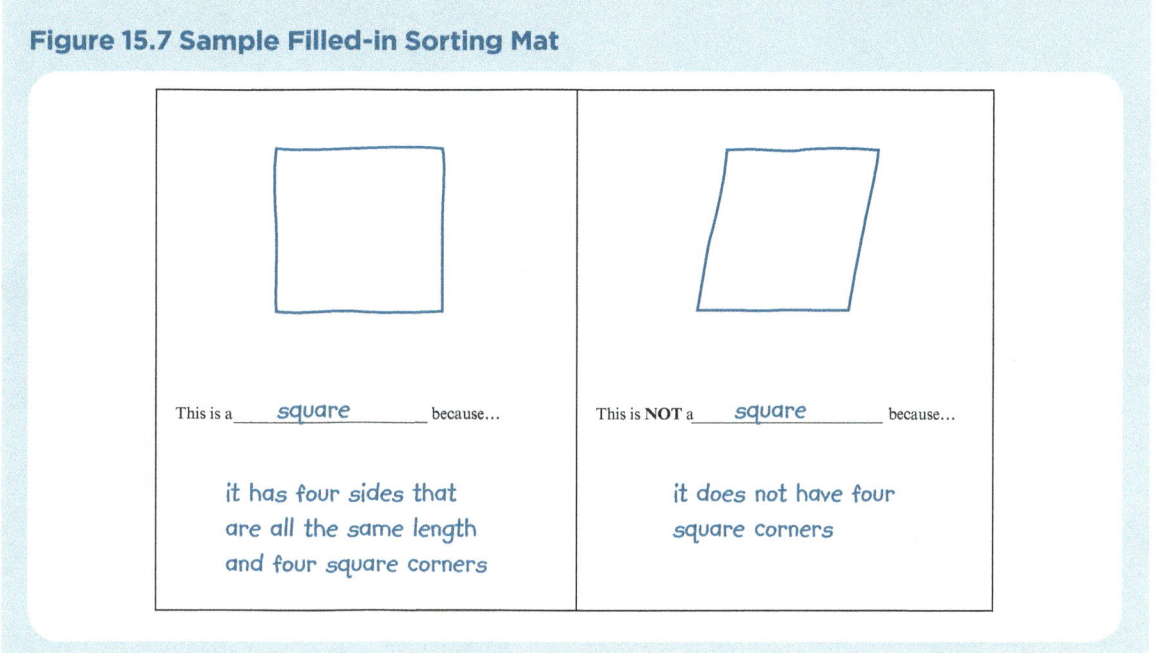

Figure 15.7 Sample Filled-in Sorting Mat

This is a ___*square*___ because…

it has four sides that
are all the same length
and four square corners

This is **NOT** a ___*square*___ because…

it does not have four
square corners

FACILITATE

1. Together with students, generate a list of the two-dimensional shapes they could draw. Write the list on the board where students can easily see it.

2. Organize students into heterogeneous pairs. Give each pair at least two student pages.

3. Encourage students to take turns:

   » The first partner will draw a shape and complete the sentence frame, "This is a ＿＿ because . . ."

   » The second partner then draws a shape that is *not* the shape drawn by the first partner and completes the sentence frame, "This is *not* a ＿＿ because . . ."

   » Partners trade roles and the second partner gets to pick a shape.

4. **Show Me/Interview.** As students draw shapes that do and do not have defining attributes, circulate through the class asking questions to push on students' knowledge of defining attributes for each shape. For example:

   » Show me how you know the sides are all the same length.

   » Show me where there is an extra side in the figure that is *not* a triangle.

» **Interview.** I noticed you didn't say that this is a rectangle because it's (small*). Why not? Can you show me an example of a shape that is (small*) but isn't a rectangle?

*name a non-defining attribute of the student's figure

Note: Consider using the Interview (individual or small group) and Show Me tools for recording responses from individual students or pairs (see Appendix B).

CLOSE: MAKE THE MATH VISIBLE

1. Group student pages according to the shape drawn on the left side.

2. Conduct a Something Similar and Something Different Gallery Walk.

3. Select and sequence student thinking for the class discussion based on the Show Me and Interviews conducted during the Facilitate phase of the task lesson. Examples of ideas to highlight:

   » Squares and rectangles both have four sides and four square corners, but squares *also* have the defining attribute of all sides are the same length.

   » Circles have no straight edges.

   » Triangles have exactly three sides.

   » Side length is not a defining attribute for "triangle."

4. If there are any student pages with incomplete reasoning, take this opportunity to have students practice drawing some counterexamples. For example (given the student reasoning in the image shown in Figure 15.8), ask the class, "Can you draw a shape with four sides that is *not* a rectangle?"

Figure 15.8 Student Reasoning Sorting Mat

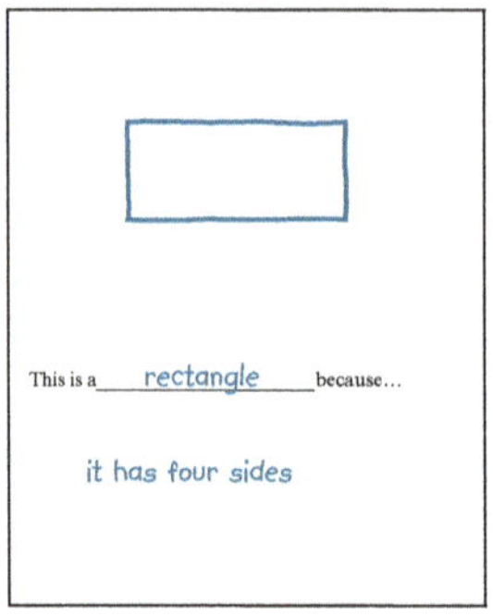

This is a ___rectangle___ because…

it has four sides

5. Ask the group who owns this reasoning how they might revise their reasoning to avoid non-rectangle four-sided figures drawn by the class.

PRODUCTIVE STRUGGLE

Normalizing this type of revision work by discussing counterexamples as a regular part of math class will help students begin to think of revising as a natural and valuable element of strong mathematical thinking.

TASK 50: WHAT IT IS AND WHAT IT'S NOT SORTING MAT STUDENT PAGE

 To download printable resources for this task, visit **resources.corwin.com/ ClassroomReadyMath/K-1**

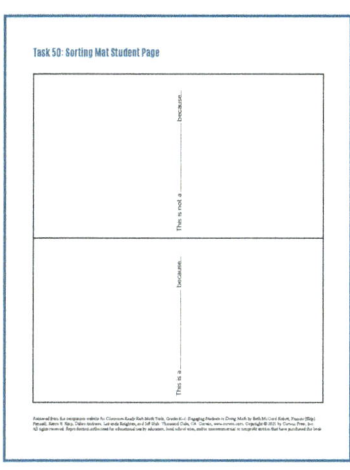

POST-TASK NOTES: REFLECTION & NEXT STEPS

Mathematics Standard

- Correctly name shapes regardless of their orientations or overall size.

Mathematical Practices

- Construct viable arguments and critique the reasoning of others.
- Look for and make use of structure.

Vocabulary

- triangle
- tilt
- defining attribute

Materials

- Triangle Sketches student page, one per group
- other tools for modeling triangles (pipe cleaners, straws, toothpicks, etc.) as desired
- marker boards and markers or sticky notes and pencils

Task 51
The *Tilted* Triangle

Recognize that triangles come in many different sizes and orientations

TASK

The *Tilted* Triangle

Mr. Booher asked the class to draw a triangle. Sonja noticed that Lorenzo's sketch looked different from hers (see Figure 15.9). She said, "Lorenzo, I don't think you drew a triangle. Yours is tilted!" Lorenzo said, "I think triangles can be tilted."

Figure 15.9 Sonja's and Lorenzo's Sketches

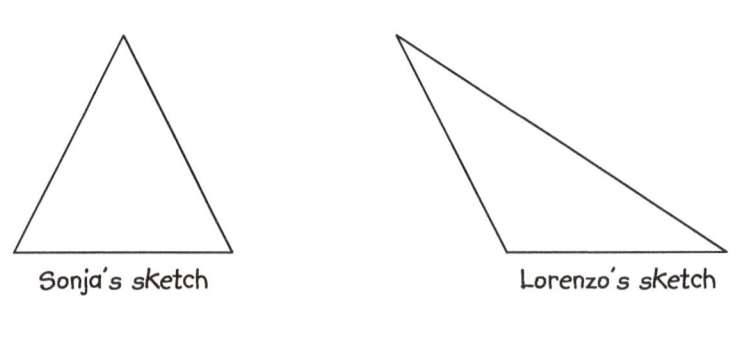

Sonja's sketch   Lorenzo's sketch

How could Lorenzo prove that his sketch shows a triangle?

TASK PREPARATION

- Consider how you will engage students in exploring the shapes in the task. Will you provide scissors so the figures can be cut out? Or perhaps students could be given straws or pipe cleaners to construct the shapes.

- Plan for students to work in small heterogeneous groups of two or three.

LAUNCH

1. Bring the class together and display only the picture from the task (Figure 15.9). Facilitate a See, Think, and Wonder.

2. Ask, "What do you notice about these two sketches? What attributes are the same? What attributes are different?"

3. Record students' ideas on the board.

4. Tell students the story of the task.

5. Have students *Turn and Talk* with a partner about Sonja's and Lorenzo's statements.

FACILITATE

1. Organize students into heterogeneous groups of two or three.

2. Provide each group with the Triangle Sketches student page and any additional tools (pipe cleaners, straws, toothpicks, etc.) as desired.

3. Ask students to work with their group to come up with a way Lorenzo could prove his sketch shows a triangle.

4. As students talk in small groups, circulate to **Observe.** Listen carefully to the arguments that students are making about the figures. Students' conceptions of the attributes of a triangle will be revealed in their discussions.

   » Are students confident that as long as the shape is closed and composed of three straight sides, it must be a triangle?

   » Do students need to rotate the triangle so it "points up" before they believe it is a triangle?

   » Do students agree with Sonja's thinking?

Note: Consider using the Observation tool to record discussion comments (see Appendix B).

5. When most student groups are ready to defend Lorenzo's argument, bring the class back together.

6. **Hinge Question/Show Me.** How could you draw a triangle that is different from Sonja's and Lorenzo's triangles? Show me. Make a sketch of a triangle that is different from both of the ones Sonja and Lorenzo drew.

7. Collect the students' sketches to use in the Close, step 2.

CLOSE: MAKE THE MATH VISIBLE

1. *Select and Sequence* groups to share their ideas about Lorenzo's triangle based on observations made during the Facilitate phase of the lesson. Highlight a variety of ideas:

   » Lorenzo's figure is closed and has three sides. Triangles have three sides, so Lorenzo's figure is a triangle.

   » Lorenzo's figure is made up of three straight edges.

   » We can rotate Lorenzo's figure to "point up" so it looks more like Sonja's triangle.

   » We could also rotate Sonja's triangle so it is "tilted."

> **! PRODUCTIVE STRUGGLE**
>
> Be sure to encourage students to share their ideas in their own words. Students are just developing their capacity for mathematical argumentation, so it's important they have opportunities to share even beginning or newly formed ideas. Ask questions as needed to clarify, but resist the urge to always restate students' ideas in your words.

2. As students share their ideas, use the triangles students drew in response to the Hinge Question (Facilitate, step 7) as examples to support their thinking.

» If students have agreed that triangles always have three straight sides, display the student-drawn triangles and ask, "Did we all draw figures with three straight sides?"

» If students are unsure whether orientation matters (e.g., does the triangle always have to sit on a flat side and point up?), rotate several of the student-drawn triangles and ask, "Is this still a triangle?"

TASK 51: TRIANGLE SKETCHES STUDENT PAGE

 To download printable resources for this task, visit **resources.corwin.com/ ClassroomReadyMath/K-1**

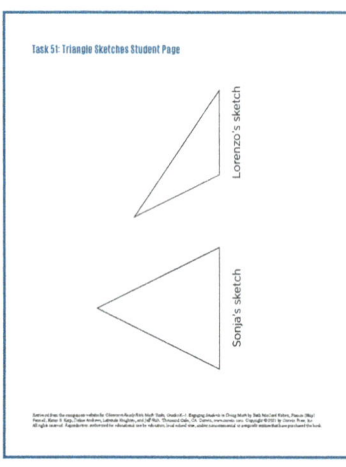

POST-TASK NOTES: REFLECTION & NEXT STEPS

Task 52

Flat or Not Flat?

Sorting two- and three-dimensional shapes

TASK

Flat or Not Flat?

Ms. Stolp made two collections of shapes (Figure 15.10). Edla thinks she knows the rule Ms. Stolp used to sort the collections. What is the rule? See if you can sort figures using the same rule. What name could you give to each group?

Figure 15.10 Two Shape Collections

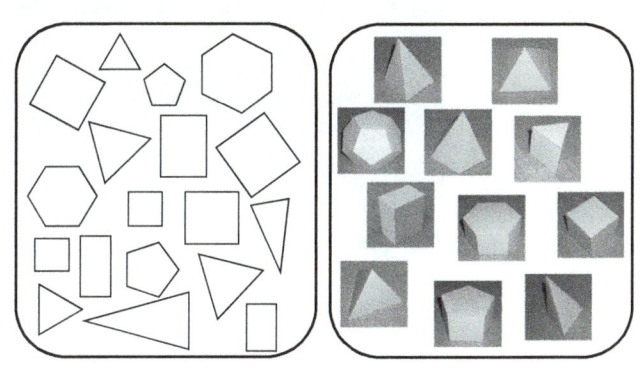

TASK PREPARATION

- Consider how you will group students for the task work.

  » During the Launch, have students share with a partner or small group before allowing ideas to be shared out to the whole class. Think about how you will arrange students for this part. Will they come up to the carpet? How will they know who their discussion partner(s) will be?

  » During the Facilitate phase of the task, students could be organized into small heterogeneous groups of three or four.

- Prepare a sufficient supply of two- and three-dimensional shapes such that each group has several shapes from each category.

  » Be sure to use printable, paper two-dimensional shapes. For this task, do not use attribute blocks or manipulative geometric shapes as they are not truly flat.

  » Assemble the three-dimensional shapes (from Nets) in advance.

Mathematics Standard

- Identify shapes as two-dimensional (lying in a plane, "flat") or three-dimensional ("solid").

Mathematical Practice

- Construct viable arguments and critique the reasoning of others.

Vocabulary

- two-dimensional shapes
- three-dimensional shapes
- flat
- solid
- line
- rotate
- length
- width
- height

Materials

- Flat Shapes student pages
- Nets student pages (assembled in advance)
- Sorting Mat student page, one per group

ACCESS AND EQUITY

Having three-dimensional shapes constructed in advance will eliminate potential barriers to participation for students and ensures adequate time for students to have opportunities to identify and compare attributes of the figures.

LAUNCH

1. Create two collections of shapes using some, but not all, of the two- and three-dimensional figures that have been prepared (see sample images in Figure 15.10).

2. Display the collections for the class. Share the story and facilitate a Notice and Wonder.

3. Have students *Turn and Talk* before sharing ideas with the class.

4. Record students' ideas on the board.

ACCESS AND EQUITY

Small-group or partner talk before whole-class sharing helps to ensure that every student has an opportunity to think before the whole-group discussion and gives the teacher an opportunity to circulate and listen for two or three ideas to highlight with the whole group. Ensure that you are pulling ideas from a wide representation of students. Students should be allowed to share their own idea or an idea they heard from their group.

FACILITATE

1. Organize students into heterogeneous groups of three or four.

2. Distribute several shapes (some two-dimensional and some three-dimensional) to each group.

3. Ask students to sort their collections using the same rule as Ms. Stolp used.

4. **Observe.** As students work, circulate through the groups and note how students are determining which shape belongs in which category.

5. **Interview.** Ask each group:

   » What could you call your two groups? Why?

   » How did you decide to place this figure (indicate a shape) in this group?

   » How is this shape the same as other shapes in this group? What attributes do they share?

   » How are these shapes (indicate one group) different from these shapes (indicate the other group)? How are their attributes different?

Note: Consider using the Observation and Interview (small group) tools for recording student responses (see Appendix B).

CLOSE: MAKE THE MATH VISIBLE

1. Once each group has sorted their small collection, facilitate a Notice and Wonder Gallery Walk.

2. Bring the class back together to share their observations.

3. *Select and Sequence* small groups to share their thinking on some of the interview questions asked during the Facilitate phase of the task. Highlight the following ideas:

   » All of the shapes in one group are flat.

     » These are called two-dimensional shapes.

   » The shapes in the other group are not flat.

     » These are called three-dimensional shapes.

     » These are called solid figures.

   » If you hold a shape from the "flat" group level with your eyes, all you can see is a line!

Figure 15.11 Demonstrating a "Flat" Shape

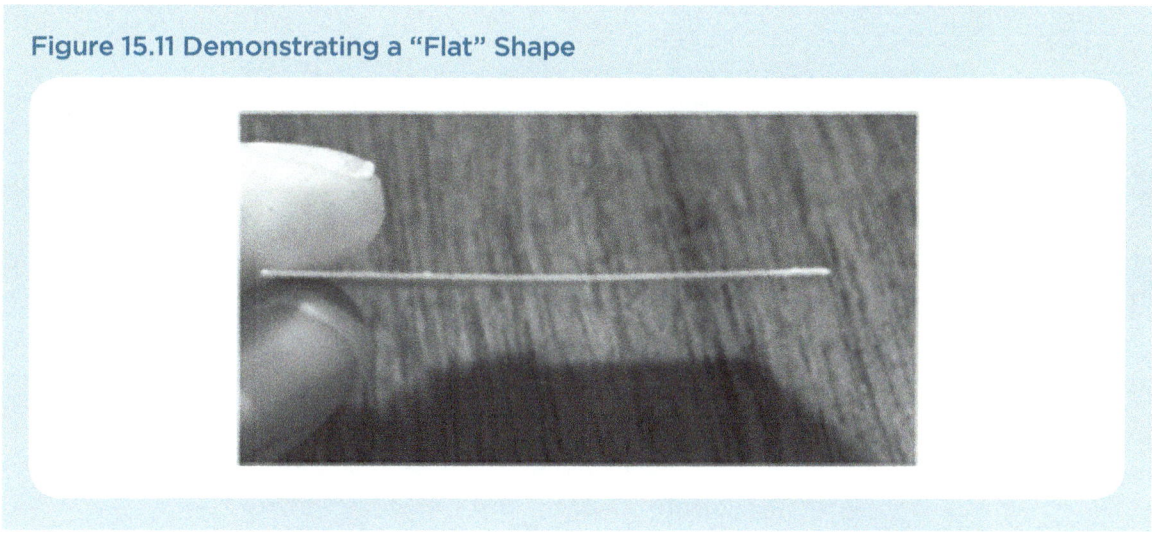

   » Encourage students to try this (as shown in Figure 15.11) using flat shapes from their own collections.

4. **Hinge Question.** Ask, "What kinds of shapes do you see when you rotate a shape from the "not flat" group?"

   » We can rotate a shape from the "not flat" group to see some of the shapes from the "flat" group. For example, we can see a square on each face of a cube (Figure 15.12).

Figure 15.12 Demonstrating a "Not Flat" Shape

- » Encourage students to try this using three-dimensional shapes from their own collections.

- » If we open up one face of a shape from the "not flat" group, we could fill up the three-dimensional shape with something like rice or beads.

  - » Encourage students to test this out using shapes from their own collections (or shapes you've prepared in advance with one open side).

TASK 52: FLAT OR NOT FLAT STUDENT PAGES

online resources ↘ To download printable resources for this task, visit **resources.corwin.com/ClassroomReadyMath/K-1**

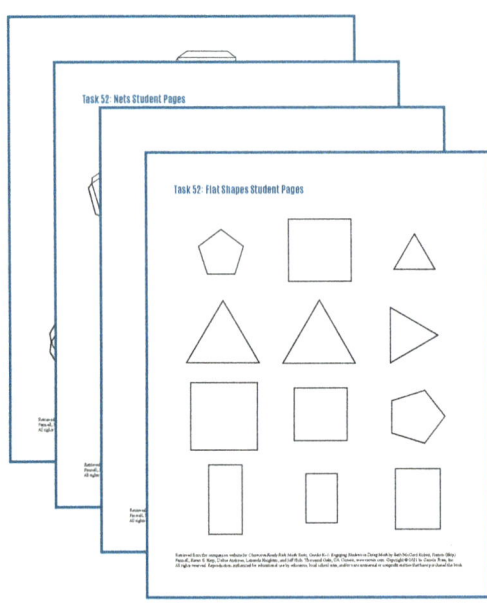

POST-TASK NOTES: REFLECTION & NEXT STEPS

Geometry
Composing Shapes and Equal Shares

TASK 53: GRADE 1: WHAT'S MY ATTRIBUTE?

Distinguish between defining attributes (e.g., triangles are closed and three-sided) versus non-defining attributes (e.g., color, orientation, overall size); build and draw shapes to possess defining attributes.

TASK 54: KINDERGARTEN: THE TRIANGLES HAVE IT

Compose simple shapes to form larger shapes. *For example, "Can you join these two triangles with full sides touching to make a rectangle?"*

TASK 55: GRADE 1: SHAPE CITY

Compose two-dimensional shapes (rectangles, squares, trapezoids, triangles, half-circles, and quarter-circles) or three-dimensional shapes (cubes, right rectangular prisms, right circular cones, and right circular cylinders) to create a composite shape, and compose new shapes from the composite shape.

TASK 56: GRADE 1: BAKE SALE

Partition circles and rectangles into two and four equal shares, describe the shares using the words *halves, fourths,* and *quarters,* and use the phrases *half of, fourth of,* and *quarter of.* Describe the whole as two of, or four of the shares. Understand for these examples that decomposing into more equal shares creates smaller shares.

Anticipating Student Thinking: This chapter, the final task chapter, extends your students' experiences in geometry. The chapter's kindergarten task engages students in using pattern blocks to create other shapes. The chapter's first-grade tasks engage students in sorting shapes based on attributes, working with three-dimensional shapes, and partitioning shapes into equal shares and defining these shares as fractions (halves and fourths). As you plan for implementing the unit's tasks, make sure to allot the time necessary for assembly and distribution of each task's materials. Also recognize that these tasks are "vocabulary heavy." That is, all of the tasks include geometry-related vocabulary that is most likely new for many of your students. As you observe, interview, and pose task Hinge Questions, monitor student use of the task's vocabulary, providing assistance as needed. Also, pay particular attention to the related expectations in the Bake Sale task, which is about partitioning shapes but connects directly to fractional parts of the shapes—an important introduction to fractions—within the context of geometry.

THINK ABOUT IT

As noted, this chapter's tasks include many geometry-specific vocabulary words. As you communicate with family members, particularly if you are teaching online or virtually, make sure to send pictures or perhaps physical links to the many vocabulary words and related concepts their children will experience in each of the task lessons.

Mathematics Standard

- Distinguish between defining attributes (e.g., triangles are closed and three-sided) versus non-defining attributes (e.g., color, orientation, overall size); build and draw shapes to possess defining attributes.

Mathematical Practice

- Look for and make use of structure.

Vocabulary

- corners
- vertices
- closed
- flat
- figure
- sides
- attribute
- defining attribute
- rectangle
- square
- square corner
- triangle

Task 53
What's My Attribute?

Reason with shapes and their attributes

TASK

What Am I? Riddles

Each of you has a riddle on a card that uses defining attributes to describe a shape. We are going to play a game using defining attributes to identify shapes. Read over your card and discuss the answer with your partner.

What's My Attribute?

Sheena sorted shapes into two groups (see Figure 16.1). What attributes did she use to sort the shapes? Label each group. Draw another shape that belongs in one of the groups. Why does your shape belong in that group?

Figure 16.1 Sheena's Sorting

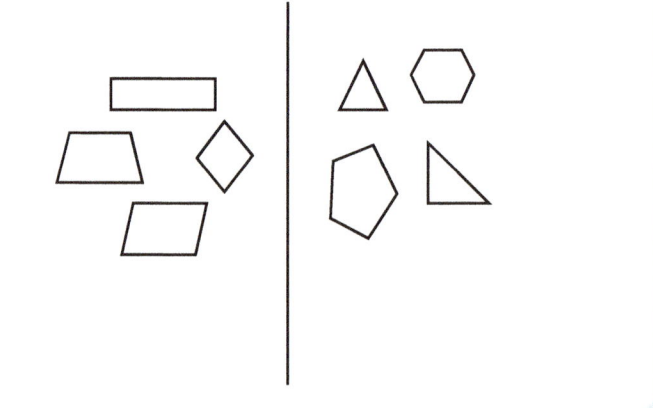

TASK PREPARATION

- The Launch activity engages students in using a game, What Am I? riddles, to activate prior knowledge in preparation for the task.

Note: Students will use the term *square corner* to describe a right angle. Introduce students to this term prior to this activity. You may want to give each student an index card to use as a "right angle finder." Chapter 15, Task 49, where students identify defining and non-defining attributes, is a good prerequisite for this task.

- Make copies of the riddle cards and cut them apart.

- Students will need to move freely around the classroom to complete the What Am I? riddle Launch activity. Determine procedures for managing student movement during this portion of the task. Select a method for placing students in pairs for the remainder of the task.

LAUNCH

1. Bring students to a whole-group discussion. Project the Figure 16.2 image or a similar image for students.

Figure 16.2 A Sample Shape

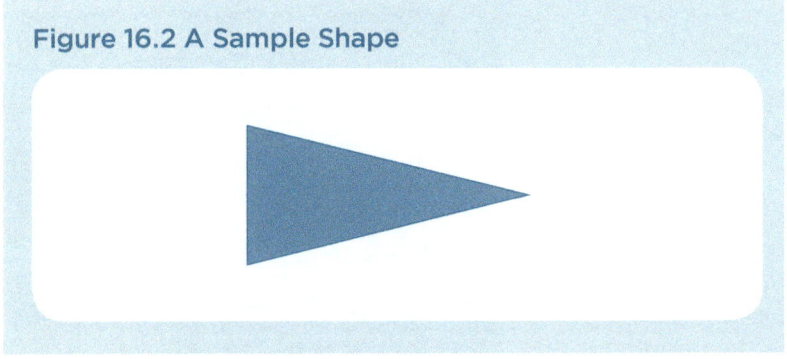

2. Ask, "How would you describe this shape?" Sample responses may include the following:

   » It has three sides.

   » It has three corners.

   » Two sides are the same length.

   » It is blue.

   » It is a triangle.

3. Record student responses. Say, "Turn and talk to a partner. Are all of these attributes true for all triangles? Which ones are true for all triangles?" Allow time for students to discuss. Students should point out that not all triangles are blue; however, having three sides and three corners is true for all triangles, so these properties are what we call defining attributes.

4. Use questioning to support student "notices" by asking questions such as, "Can you draw a triangle that does *not* have three corners? Can you draw a triangle that is *not* blue?" Later, follow up with, "What does it tell us about the attribute blue, when you can draw a triangle that is not blue?"

5. Organize students in pairs and distribute a What Am I? riddle card to each student. Say, "Each of you has a riddle on the card that uses defining attributes to describe a shape. We are going to play a game using defining attributes to identify shapes. Read over your card and discuss the answer with your partner."

- hexagon
- pentagon
- octagon
- non – as in non-three-sided shape

Materials

- **What Am I? Riddle Cards** student page, one card per student (print student page on cardstock and cut apart before distributing)
- **What's My Attribute?** student page
- pencils
- index cards

6. After each partner has shared, introduce students to the game. Students will use *Quiz-Quiz-Trade* to solve the riddles. *Quiz-Quiz-Trade* is a cooperative learning structure that promotes collaboration and shared understanding among students as they interactively explore lesson concepts.

   » Directions: Partner A reads Partner B the riddle on their card. Partner B responds, and Partner A tells them if the answer is correct. Partner B asks Partner A the riddle on their card, Partner A responds, and Partner B tells Partner A if they are correct. Partners then trade cards, find new partners, and repeat process.

   » Allow students to switch riddles at least one time so that they get to "quiz" two different partners. Allow more if time permits.

7. Circulate and **Observe** students as they work with a partner to solve the riddles. Listen to see if students need support or have questions.

8. Bring students together as a group to debrief the activity. Ask, "Were any of the riddles more difficult to figure out than the others? How did you use the defining attributes to help you identify the shape?"

9. Read each What Am I? riddle and allow students to provide the answer and their thinking. Glue each riddle to a piece of chart paper. Select a student volunteer to draw an example of each shape described next to the corresponding riddle. Where appropriate, encourage students to show more than one drawing to represent each shape. (For example, try to get students to draw multiple representations of triangles, not just equilateral triangles, and include triangles in varied orientations, such as pointing down.) Use these drawings as an anchor chart for future experiences.

ACCESS AND EQUITY

It is important to allow students to have this time to use their prior experiences and the support of a peer to unpack the riddle before engaging in the activity. Allowing students to work with a peer in a nonthreatening manner builds confidence, encourages greater participation, and results in more thoughtful discussions.

STRENGTHS SPOTTING

Highlight students who enjoy finding counterexamples and disproving ideas, which is a strength that may not be recognized. These students want to make sense of the mathematics they are learning and feel encouraged when they are supported.

FACILITATE

1. Facilitate a *Think-Pair-Share* protocol for this task.

   » **Think.** Introduce students to the task. Say, "Sheena sorted the shapes below into two groups." Show them Sheena's sorting (Figure 16.3).

Figure 16.3 Sheena's Sorting

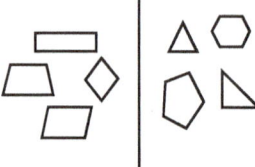

» "What attributes did she use to sort the shapes? Label each group. Draw another shape that belongs in one of the groups. Why does it belong in that group?" Allow students to work independently on the task.

» Have students share their thinking with a partner. Students discuss their responses, look for similarities and differences, and share how they used defining attributes to help them with the task.

» Have students debrief and share responses with the whole class.

2. **Observe** students as they work on the task. Watch as students examine the sorted shapes. What are they noticing about defining attributes such as number of sides, straight sides, number of corners, and so on? What non-defining attributes are they attending to such as size, orientation, color, and so on?

3. **Interview.** As students work individually and with their partners, ask questions and/or ask for responses to prompts such as the following:

» What is the same about the shapes in this group? How are the shapes in the other group different?

» How are the attributes of shapes in one group alike? How are they different?

4. When a student has identified the sorting labels, ask, "How can you prove you are correct? Why does the shape you drew belong to this group? **Show Me** how you know this shape belongs to this group."

PRODUCTIVE STRUGGLE

Asking students to identify how the shapes in a group are alike and how they are different from the other group helps students analyze the shapes. These questions can be used to help scaffold the task for students who need more support in solidifying their thinking.

Note: Consider using the Observation, Interview (small group), and Show Me tools for recording student pair or individual student responses (see Appendix B).

CLOSE: MAKE THE MATH VISIBLE

1. Bring students a whole-group discussion.. Reread the task, and allow students to share their thinking. Students should share that the group of shapes on the left are all four-sided shapes and the other group is not four-sided shapes. Allow students to share how they used the defining attributes of the shape to determine the labels.

2. Facilitate a Gallery Walk to allow students to see which shapes their peers added to the sort. Encourage students to ask questions and look for similarities and differences about the new additions.

ACCESS AND EQUITY

Students can add a shape to either group; the only requirement is that students can explain why the new shape belongs to that group using defining attributes. This promotes student choice and gives students the opportunity to examine examples and non-examples.

3. Conclude the task lesson with the following **Exit Task:** Marissa sorted shapes in the two groups shown in Figure 16.4.

Figure 16.4 Marissa's Sorting

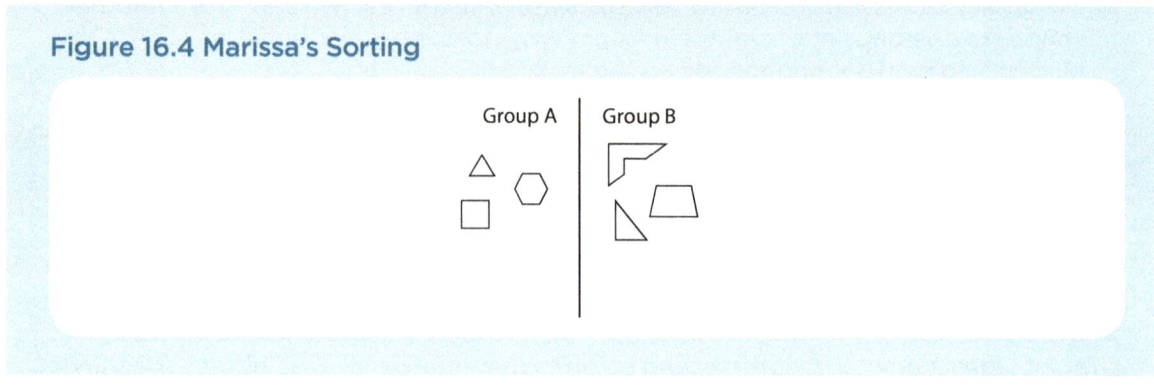

4. She added this shape [＿＿＿] to Group B. Explain why this shape does not belong in Group A.

TASK 53: WHAT'S MY ATTRIBUTE? STUDENT PAGES

online resources ⬈ To download printable resources for this task, visit **resources.corwin.com/ ClassroomReadyMath/K-1**

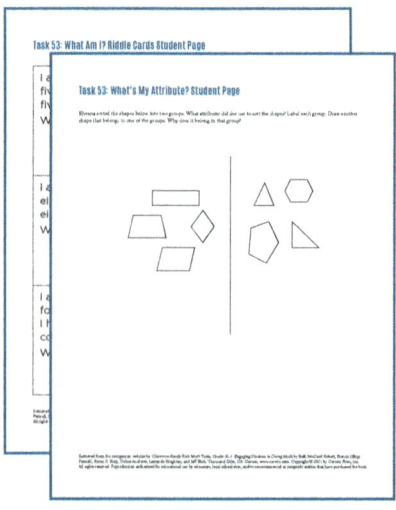

POST-TASK NOTES: REFLECTION & NEXT STEPS

Task 54

The Triangles Have It

Compose and decompose shapes

TASK

The Triangles Have It

Tyler used pattern block triangles to build larger shapes (Figure 16.5). Each of his new shapes can be made with exactly six triangles. Which shapes did Tyler make? Show how you know.

Figure 16.5 Tyler's Shapes

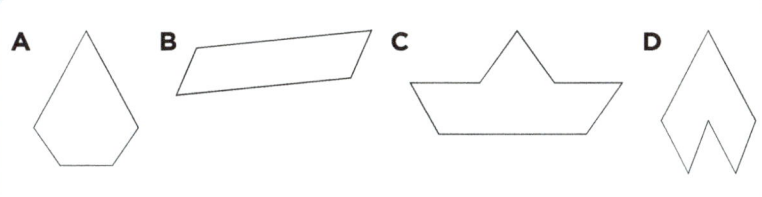

TASK PREPARATION

- Prior to this task, students should engage in activities that allow them to analyze shapes where they discuss similarities, differences, and identifying attributes.

- Prepare baggies of assorted pattern block shapes (remove the tan rhombus). Student pairs will need several of each shape.

- Duplicate the Pattern Block Spinners on heavy paper and laminate if possible. (If lamination is not an option, place the spinner templates in a sheet protector.) Have work mats available for students.

LAUNCH

1. Divide students into pairs and distribute baggies of assorted pattern blocks, pencils, and blank paper. Say, "Each of you select a shape from your baggie. Turn and talk to a partner and describe the attributes of your shapes." Allow time for students to describe their shapes to their partners. Say, "Now put the two shapes together with sides touching exactly (demonstrate if needed) to make a new shape. Trace your shape on paper and describe your new shape." Figure 16.6 shows how a combined shape might be created.

Mathematics Standard

- Compose simple shapes to form larger shapes. *For example, "Can you join these two triangles with full sides touching to make a rectangle?"*

Mathematical Practices

- Make sense of problems and persevere in solving them.

- Attend to precision.

Vocabulary

- triangle
- build
- alike
- different
- shape
- square
- trapezoid
- hexagon
- rhombus
- side
- attributes
- corner
- compose

Materials

- pattern blocks, one set per student pair
- Pattern Block Spinner student page, one per student pair
- large paper clips for spinners, one per student pair
- number cubes, one per student pair
- The Triangles Have It Recording Sheet student sheet, one per student
- work mats (blank paper, colored if possible)
- blank paper, several sheets per student
- pencils

Figure 16.6 Putting Shapes Together

  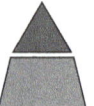

Partner A's shape + Partner B's shape =

ACCESS AND EQUITY

As there may be variations in the fine motor skills of young students, students can support each other by having one partner hold the shape while the other traces around it to show their work.

2. As students describe their pattern block shapes and the new shapes they make, listen to student conversations for mathematical language that includes the shape names, number of sides, and number of corners. Kindergarten students may be unfamiliar with the names of some of the shapes in pattern block sets; however, students should be able to describe a shape using attributes such as the number of sides, number of corners, and sides of equal length in their descriptions.

3. Distribute the Pattern Block Spinners, paper clips, and number cubes to student pairs. Say, "Now you're going to use these spinners and number cubes to play a Make That Shape game. Partner A spins the spinner to determine which pattern block shape to select. Partner B tosses the number cube to determine how many of that shape you will use to make a new shape. For example, if the spinner lands on the blue rhombus and your partner tosses the cube and gets a 4, you make a new shape using 4 rhombuses. Trace the new shape on the paper."

4. Allow students to play at least two rounds of the game so that each partner gets a turn to make a shape.

ACCESS AND EQUITY

The Launch activities allow students to review composing shapes while incorporating student choice and creativity, and, at the same time, encouraging students to actively engage in the task. These activities also allow students to manipulate the shapes using a variety of movements in preparation for the Facilitate task.

5. As students work on both of the Launch activities, circulate among student pairs as they work. Consider the following **Interview** questions:

   » What shapes did you use to make your new shape? How did you put these two shapes (triangles, squares, trapezoids, hexagons, rhombuses) together to make a new shape?

   » How many sides and corners are in your new shape?

   » How do you know it is a (rectangle, triangle, pentagon . . .)?

   » Is there another way to make this shape?

   » How is your shape the same as your partner's shape? How is your shape different from your partner's shape?

6. **Observe** students as they work. Make note of the language students use as they describe their shapes and describe how they move the shapes to compose new shapes.

   » What attributes are students using to describe the shapes? (**Note:** The term attribute is not a term assessed at the kindergarten level; however, we want to encourage students to use defining attributes that describe the characteristics that make a shape a shape. By hearing the vocabulary early, the students will build familiarity with these terms.)

   » What strategies do students use to manipulate the shapes to compose the new shape? (Encourage students to share their strategies with others, especially students who may need additional support in this area.)

7. Bring students back to whole group and allow students to share their shape pictures, discuss their observations, and share their thinking.

Note: Consider using the Observation and Interview (small group) tools for keeping track of what you observe and of student interview responses (see Appendix B).

FACILITATE

1. Introduce the task to students. Say, "Tyler used pattern block triangles to build larger shapes. Each of his new shapes can be made with exactly six triangles. Look at the shapes in Figure 16.7.

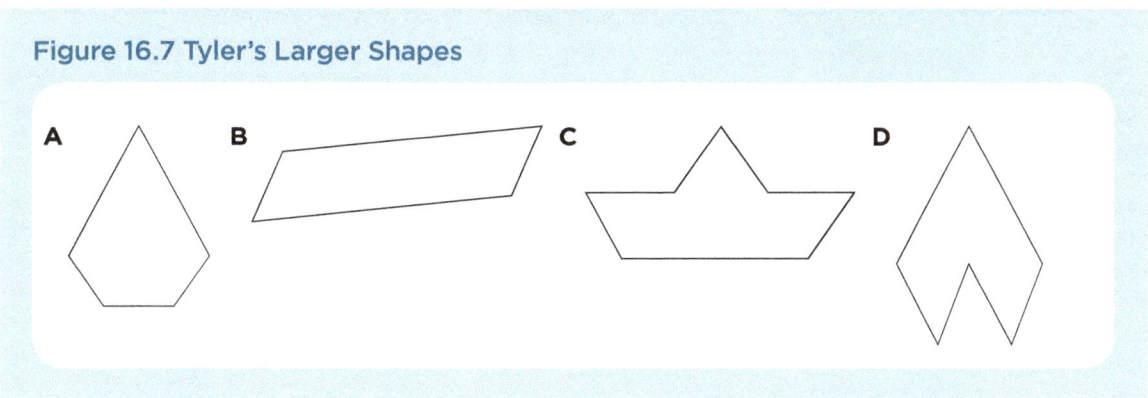

Figure 16.7 Tyler's Larger Shapes

Which shapes did Tyler make? Show how you know" (**Show Me**).

2. Distribute the The Triangles Have It recording sheet to student pairs. Allow student pairs to work on the task.

3. **Observe** the student pairs as they use trial and error to fill the shapes with triangles and figure out which shape uses more than six triangles. Also observe the students as they place the shapes on the template. What strategies do students use to fit the pattern block shapes on the template?

4. Facilitate a *Pair-to-Pair Share* with students. Allow time for students to share their thinking, ask questions, and share their strategies. Make note of strategies student pairs used to find out which shape is not made up of six triangles.

5. **Hinge Question.** How did you use what you knew about making larger shapes from smaller shapes to help you solve this task?

6. As you circulate among students, make note of student agreements and/or disagreements about the figures Tyler made. Prepare to have those students share during the upcoming whole-group discussion.

PRODUCTIVE STRUGGLE

Some students may struggle with figuring out how/when to flip or turn the pattern block shapes to cover the shape outlines. Working with a partner allows students to work together and support each other. You may also find it helpful to suggest that students try to move the blocks in different positions or orientations to make them fit.

ALTERNATE LEARNING ENVIRONMENT

To provide access to students in remote learning settings, use Google Classroom to assign a Google doc such as this example: The Triangles Have It, https://docs.google.com/presentation/d/1c7-6LrsTdm7TSUCUXE8AwVvz9ytYMyhfNG9pbENL9HY/edit?usp=sharing

CLOSE: MAKE THE MATH VISIBLE

1. Bring students back to the whole group. Ask students to show how they know how many triangles each shape used. Ask, "Which shapes did Tyler make?"

2. As students share their responses, look for student agreement and/or disagreement about which figures used six triangles.

   » If all students agree on which shapes Tyler made, tailor the discussion to focus on the strategies that students used to figure out which shapes Tyler made with the six triangles.

   » However, if there is some disagreement on which shapes Tyler made, tailor the discussion to focus on students defending their work.

3. **Exit Task.** Select shape A, B, C, or D. Show a different way to cover the shape with pattern block shapes other than with six triangles. Observe students as they complete the Exit Task. If time allows, have students share the new way to cover their shape.

TASK 54: THE TRIANGLES HAVE IT STUDENT PAGES

 To download printable resources for this task, visit **resources.corwin.com/ ClassroomReadyMath/K-1**

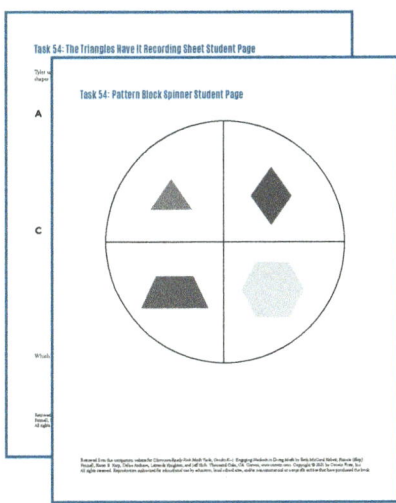

POST-TASK NOTES: REFLECTION & NEXT STEPS

Mathematics Standard

- Compose two-dimensional shapes (rectangles, squares, trapezoids, triangles, half-circles, and quarter-circles) or three-dimensional shapes (cubes, right rectangular prisms, right circular cones, and right circular cylinders) to create a composite shape, and compose new shapes from the composite shape.

Mathematical Practice

- Attend to precision.

Vocabulary

- composite shape
- solid
- face
- three-dimensional shape
- cone
- rectangular prism
- cube
- cylinder
- sphere
- above, below, on the edge

Task 55

Shape City

Compose three-dimensional shape cities

TASK

Shape City

Figure 16.8 A City

Source: mspoint/iStock.com

You and your group members are to build a Shape City (Figure 16.8) using three-dimensional shapes (solids). Your city must follow the rules provided below:

1. You may use as many pieces as you want, but you must use at least three different solids.

2. Your city must have a building that has more than one layer. This means that you must stack solids to build more than one layer.

3. Label each building with its name (Red Stick Elementary, Gumbo Park, Jamar's house, and Baton Rouge Zoo).

4. Be creative!

TASK PREPARATION

- Students do not need to learn formal names such as "right rectangular prism."

- Students should have experience with identifying attributes of two- and three-dimensional shapes prior to this task.

- Prior to introducing the task, collect recycled items that represent three-dimensional shapes. (You can involve parents and families in this process and let them know ahead of time to save the items).

- Make copies of (cut apart and assemble) the three-dimensional nets if planning to have these available for student use.

- You may also want to visit other grade levels to borrow sets of three-dimensional solids in order to have a sufficient supply for student use.

- Make plans for managing student movement during the Facilitate portion of the task.

LAUNCH

1. If available (personal copy or within your school's library/media center), read the children's literature book *Captain Invincible and the Space Shapes* to students. Discuss story details, specifically the attributes of the solid shapes Sam uses in his pretend spaceship. Record attributes on a class anchor chart.

2. Divide students into pairs and distribute sets of three-dimensional solids and a manila folder. Allow time for students to refamiliarize themselves with names and attributes of the solid shapes. Say, "I'm going to combine solid shapes to build a composite shape behind this folder. You and your partner will listen as I describe my shape and then try to build the same shape with your materials. Build a shape using the solids and shield it from other students seeing it by using the manila folder."

   » "My first shape is a rectangular prism with the smallest face touching the table. I put a cube on top of the prism. Then I put a cone on top of the cube so that the flat face is touching the top of the cube. Build my shape."

   » "The first shape in my building is a rectangular prism with the largest face touching the table. I put a cylinder on top of the right end of the prism. I put another cylinder on top of the left end of the prism. Then I put another rectangular prism on top of both cylinders like a bridge. Build my shape."

3. **Observe** students as they build the composite shape. Are students able to select the correct three-dimensional shapes and use the descriptions to place each shape in the correct orientation or position?

4. Tell students that they will get to play the mystery shape game with a partner. Partner A builds a shape with geometric solids while hiding their shape behind the manila folder. Then they describe the shape to Partner B as Partner B builds the composite shape. Partner B checks their work. Then partners trade roles and Partner B builds and describes a shape for Partner A to build.

5. **Observe.** Circulate among student pairs and observe students as they work. Are students using precise language to describe the position and orientation of the solid shapes in their discussions?

Materials

- literature book: *Captain Invincible and the Space Shapes* by Stuart J. Murphy

- three-dimensional solids, one set per group of two to four students

- manila folders, one per pair of students and one for the teacher

- recycled items to represent three-dimensional solids (cereal boxes, food boxes, food cans, makeup containers, empty toilet paper or paper towel rolls, party hats, cake mix boxes, glue sticks)

- three-dimensional nets (from Task 52) if you do not have access to three-dimensional shapes

- Shape City Recording Sheet student page

- Shape City Building Shapes Table student page (use one for each building)

- tape

- glue sticks

- poster paper (see Shape City Sample Poster student page)

6. **Hinge Point Question** (ask students before the reveal). How do you know you have the same composite shape your partner described? Use math vocabulary to support your answer.

7. Add any new vocabulary words generated to the class anchor chart.

FACILITATE

1. Have each pair of students join another pair to make groups of four. Distribute the Shape City recording sheet and materials. Allow time for students to examine the materials and identify similarities to the wooden, plastic, or paper solids.

Note: It would be helpful to place the building materials in a centralized spot and allow students to freely select the desired materials for their cities as needed.

STRENGTHS SPOTTING

Empowering students with the opportunity to control selection and distribution of materials for this task could provide an opportunity for students to demonstrate leadership and decision-making strengths that might otherwise be missed in a more controlled environment.

2. Introduce students to the task. Say, "You and your group members are to build a Shape City using three-dimensional solids. Your city must follow these rules:

   i. You may use as many pieces as you want, but you must use at least three different solids.

   ii. Your city must have a building that has more than one layer. This means that you must stack solids to build more than one layer. (To provide students with a context for this rule, refer back to the composite shapes in the Launch activity where the first shape had two layers and the second one had three layers.)

   iii. Label each building with its name (Red Stick Elementary, Gumbo Park, Jamar's house, and Baton Rouge Zoo).

   iv. Be creative!

3. **Observe** students as they build their cities. Make note of the language students use to talk about their buildings and the shapes. How are students using what they know about the attributes of the solids to make decisions about placement and position of shapes?

4. **Interview.** Consider the following interview questions as you visit the student groups:

   » Which solid shapes did you choose to use? Why?

   » Were there any solids you chose not to use? Tell why.

   » What's the same about your composite shapes? What's different about them?

   » Which shapes can you stack another shape on top of? Which ones can't have a shape stacked on top? (treat stackable as an attribute)

5. Allow students to place their Shape Cities on the poster paper to make it easy to display.

6. Facilitate a *One Stay, the Other Stray* protocol with students. Each group selects one person to remain with their Shape City to answer questions and the other group members visit another group to examine their work. Encourage students to use the list of three-dimensional shapes on the Shape City recording sheet to guide their questions to their classmates and to check their classmates' Shape Cities to make sure they meet

all of the criteria. When students return to their groups, allow time for them to debrief and make any adjustments to their Shape City recording sheet if needed.

Note: Consider using the Observation and Interview (small group) tools for organizing and recording student observation and interview responses (see Appendix B).

CLOSE: MAKE THE MATH VISIBLE

1. Bring students back to the whole group. Allow students to participate in a Gallery Walk so that they can see all of the cities.

2. Ask the students, "What shapes were used more often than other shapes? What attributes do you think made those shapes more popular than the others?"

3. **Hinge Question.** What shapes were used less often than others? Tell why you think these shapes were used less often.

4. Find groups that used the same solid shapes to build different composite shapes. Ask the students to describe how their Shape Cities are the same and how they are different. If possible, keep these on display to share with your school community.

TASK 55: SHAPE CITY STUDENT PAGES

 To download printable resources for this task, visit **resources.corwin.com/ClassroomReadyMath/K-1**

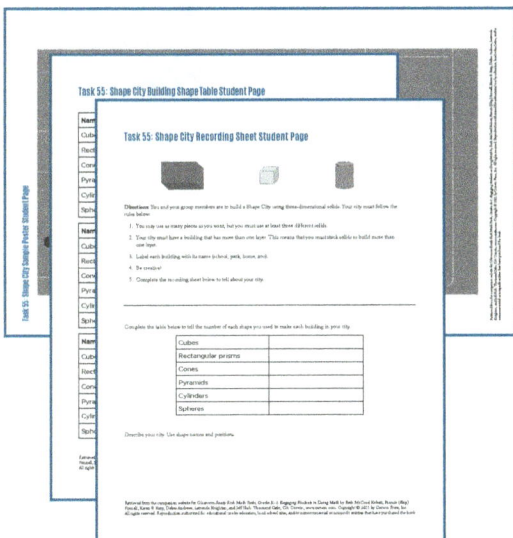

POST-TASK NOTES: REFLECTION & NEXT STEPS

Mathematics Standard

- Partition circles and rectangles into two and four equal shares, describe the shares using the words *halves, fourths,* and *quarters,* and use the phrases *half of, fourth of,* and *quarter of.* Describe the whole as two of, or four of the shares. Understand for these examples that decomposing into more equal shares creates smaller shares.

Mathematical Practices

- Construct viable arguments and critique the reasoning of others.
- Model with mathematics.

Vocabulary

- half, halves
- fourth, fourths
- partition
- fraction

Task 56
Bake Sale

Partition shapes into halves and fourths

TASK

Bake Sale

Brandon and Kenya both used white and blue frosting (icing) to decorate cakes for the bake sale. Figure 16.9 shows how they decorated their cakes. Brandon said that he needs more white frosting than Kenya. Kenya said that they both need the same amount. Do you agree with Kenya or Brandon? Show how you know.

Figure 16.9 Brandon's and Kenya's Cakes

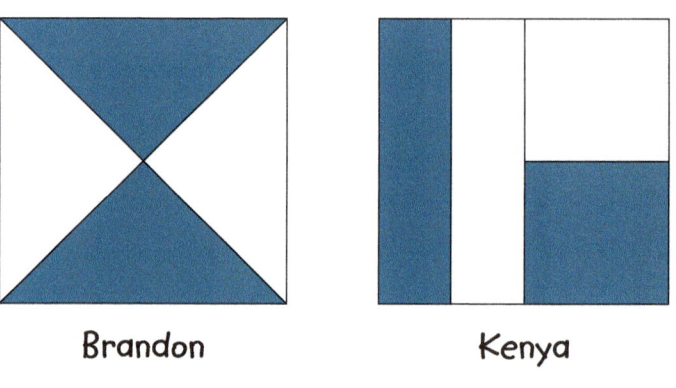

Brandon   Kenya

TASK PREPARATION

- Prior to engaging in this task, students should have experience with partitioning shapes in fourths in multiple ways and comparing fourths of the same shape.

- Make copies of the paper squares reproducible prior to the task.

- Pre-cut shapes for the Launch activity.

- Assemble materials in plastic baggies for easy distribution or create a station area in the classroom for easy pickup as students engage in the task.

LAUNCH

1. Display the Which One Doesn't Belong? prompt shown in Figure 16.10. (Have paper, crayons, and/or pencils available for student use.)

Figure 16.10 Which One Doesn't Belong?

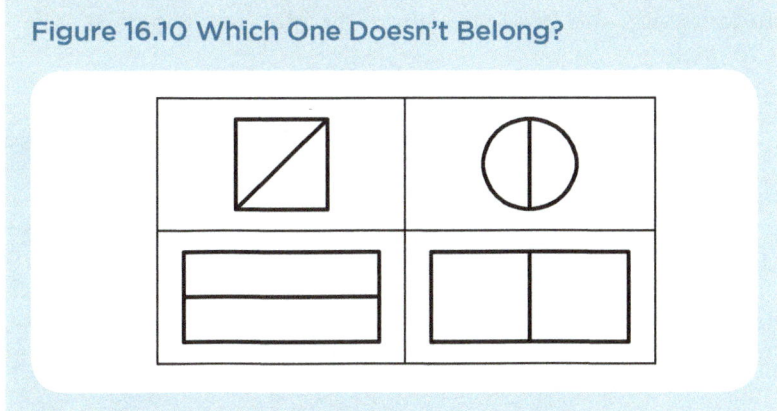

2. Ask, "Which one doesn't belong? You have about two minutes of individual time before working with a partner. Be prepared to explain which one doesn't belong and tell your reasons why. You may use paper and crayons to help with your thinking." After about two minutes, allow students to share their thinking with a partner.

3. Allow time for student pairs to share their thinking. Encourage students to use a variety of methods such as mathematical reasoning, pictures, or words to prove their answer makes sense.

4. Bring the class together for whole-group discussion. Encourage students to explain their reasoning for why one of the choices doesn't belong. Record student responses. Students should conclude that each choice shows a shape partitioned into halves. Allow time for students to review what it means to partition a shape into halves. (If needed, refer to class anchor charts.) Discuss the different ways half is shown within the shapes: circle partitioned horizontally, square partitioned diagonally, rectangle partitioned horizontally and vertically. Record any new student ideas on chart paper.

5. Note the following: "We just looked at shapes that were partitioned (split) into halves. Can someone tell me how we describe shapes we partition into four equal pieces?" Allow time for students to respond. (Students should say "fourths" for the shape description.) Show students a large piece of construction paper and ask a student volunteer to show the class how to partition the paper into fourths.

6. Divide students into pairs and distribute the Show Halves, Show Fourths Spinner; paper clips; pencils; and pre-cut shapes.

7. Select a student volunteer to be your partner to model the game procedures. Allow time for students to play the game.

Materials

- paper squares, one copy per student
- scissors, one pair per student
- crayons, one set per pair
- pencils, one per student
- grid paper, one copy per student
- pre-cut shapes on construction paper or cardstock, one set per pair
- Shapes for Launch student page
- Cake Squares student page
- Bake Sale Recording Sheet student page, one per student. **Note:** It is not expected that teachers will have access to colored ink to print this page for all students. However, you could print the document in black and white and have students color in the parts of the cakes as directed.
- Show Halves, Show Fourths Spinner student page, one per pair of students
- paper clips, one per pair of students

8. **Observe** students as they play the game. Can students easily show halves and fourths? What strategies do students use to find halves and fourths (folding, drawing lines, comparing areas—cutting apart shapes and laying the pieces on top of each other to ensure they are the same size, etc.)?

9. **Interview.** As you visit the student pairs and observe, randomly interview the pairs, asking questions such as the following:

   » What do the words halves and fourths mean?

   » How do you know your parts are halves? Fourths? How could you prove that you are correct?

   » Do halves of one whole shape have to be the same size? Do halves have to be the same shape?

   » Can you find a different way to show halves/fourths of this shape?

   » What can you do to make sure you have made equal parts of the rectangle, square, circle, hexagon?

ACCESS AND EQUITY

The Launch activity allows students to practice and solidify their understanding about partitioning shapes in halves and fourths. Students can also build their confidence on the topic while allowing the teacher to collect formative assessment data to help inform instruction.

10. To close the Launch activity, allow students to share their findings and responses to the Show Halves, Show Fourths activity. Add any new student thinking to the class anchor charts.

FACILITATE

1. Distribute Bake Sale recording sheets and materials (paper squares, scissors, pencils or markers, and/or grid paper) to students.

STRENGTHS SPOTTING

In addition to allowing students to use varied representations to solve a mathematical task, this activity allows students to use reasoning skills to solve the task. When students evaluate Brandon and Kenya's thinking, they recognize that making mistakes and revising ideas is part of the learning process.

2. Introduce the students to the task. Say, "Brandon and Kenya both used white and blue frosting to decorate cakes for the bake sale. The picture below shows how they decorated their cakes." Ask, "What do you notice, wonder? Turn and talk to a partner."

3. Select a few student pairs to share their observations. Say, "Brandon said that he needs more white frosting than Kenya. Kenya said that they both need the same amount. Do you agree with Kenya or Brandon? Show how you know."

4. **Observe** students as they work, using the following **Show Me** question: "How can you use what you know about partitioning fourths to show me how you know the amount of frosting needed?"

5. Once students have had a chance to solve the task, divide the students into pairs. Allow students to share their work with a partner. Then, ask the following guiding questions:

   » Did you and your partner have the same answer?

   » Did you use the same or a different method to find your answer?

» Explain to your partner how you know your answer is correct. Can you show your answer a different way?

» Do halves of the same whole have to be the same size? Do halves have to be the same shape?

6. **Observe** and make note of student solution strategies. Select student volunteers who used different methods to share during whole-group/reflection time.

Note: Consider using the Observation, Interview (small group), and Show Me tools for recording responses as you observe, interview, and use the Show Me technique in the Launch and Facilitate portions of this task lesson.

CLOSE: MAKE THE MATH VISIBLE

1. Bring students to circle time. Allow time for the students to share their findings from their partner discussions.

2. Ask students, "Do you agree with Kenya or Brandon?" Students should share that both Kenya and Brandon partitioned their cakes in fourths and colored two pieces white and two pieces blue. So, they both used the same amount of white frosting and Kenya was correct.

3. Allow selected student volunteers to share their thinking and solutions. Place emphasis on students using what they know about partitioning shapes in fourths to prove how they know their answer. This should include encouraging students to use cutting and/or folding to prove that the parts in Kenya's cake all represent equal areas. Listen carefully to student reasoning.

4. Add any new student understandings to the class anchor chart. It is important to highlight the idea that an equal share does not have to be the same shape (congruent) but must have an equal area.

TASK 56: BAKE SALE STUDENT PAGES

online resources ⟶ To download printable resources for this task, visit **resources.corwin.com/ ClassroomReadyMath/K-1**

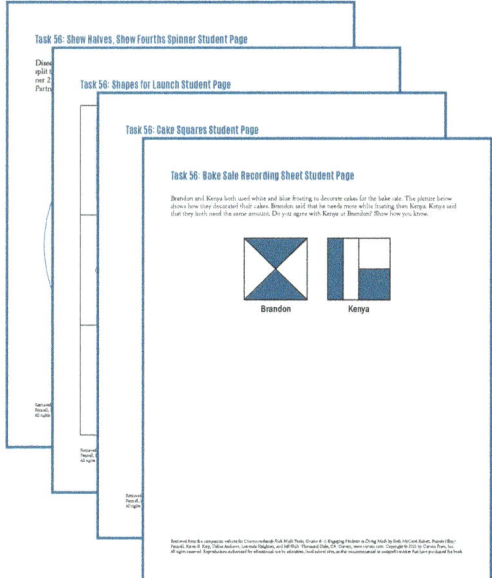

Your Turn

Now that you have had an opportunity to implement lots of *doing-math tasks*, we imagine that you have clear ideas about what works well in your classroom and how these tasks have specific characteristics that truly engage your students in mathematical thinking. We admire and respect the ways that teachers innovate and create to engage their students in meaningful mathematics learning experiences and want to provide support to you as you search for, adapt, and create more tasks with these features to address your students' specific learning strengths and needs.

So where do your mathematics tasks come from now, beyond those in this book? As you consider teaching an upcoming mathematics standard, thinking about and searching for rich tasks to engage your students in *doing-math* within the context of the lesson becomes an obvious and important concern and a daily element of your preparation. Here's what we know. The curriculum materials that you currently use—your textbook, trusted online sources, the professional development sessions you've attended and/or supplemental materials you have acquired over the years—all represent *possible* starting points, as does your own creativity in actual task creation. As you consider the tasks that you encounter in curriculum resources, or have gathered through other means, you may naturally want to adapt the tasks to meet your students' needs. This chapter will support you in selecting, adapting, and creating *doing-math tasks*. The chapter also includes questions that teachers often ask when they do this work, as well as an exploration of the responses to those questions.

Your Turn: Selecting Tasks

In Chapter 1, we explored important characteristics of *doing-math tasks* and their implications for creating high-cognitive-demand instruction. Now that you have had the opportunity to implement a collection of tasks, and have likely developed some of your own necessary criteria, you can apply the research-informed decision points introduced in Chapter 1 (pp. 5–7) to evaluate tasks that you would like to implement in your classroom. Let's take a look at four tasks that Carlis, a first-grade teacher, found in her district-supplied curriculum resources.

Figure 17.1 Example: First Grade

Content Standard: Work with addition and subtraction equations.

Understand the meaning of the equal sign, and determine if equations involving addition and subtraction are true or false.

Mathematical Practices: Reason abstractly and quantitatively; attend to precision.

Task A

Is the equation true or false? Prove how you know with counters or sketches.

$$9 + 6 = 16$$
$$3 = 8 - 6$$
$$14 - 6 = 8$$
$$4 + 4 = 10$$

Task B

Lindy and Talaya were collecting rocks for their rock collections. Lindy had 14 rocks but left 5 of them at her grandmother's house. Talaya had 2 rocks and found 7 more at the park. Who has more rocks, Lindy or Talaya?

Task C

Arnie and Farrah played a game. They rolled two dice and added the numbers to find the sum. Each time they rolled the dice, they recorded their totals. The person who rolled the higher total won the round.

Arnie	Farrah
6 + 5 =	3 + 6 =
3 + 4 =	5 + 2 =
5 + 4 =	3 + 3 =
1 + 5 =	3 + 3 =
4 + 6 =	5 + 5 =

Task D

Write the number to make the equation true.

$$3 + \square = 2 + 7$$
$$5 + 9 = 8 + \square$$
$$4 + 6 = \square + 2$$

Next, let's take a look at how Carlis used the research-informed task selection decision points to record notes about each task and use that information to make a decision about the task she would implement in her own classroom.

Figure 17.2 Task Selection Decision Points

Task Selection Decision Points	Task A	Task B	Task C	Task D
Does the task connect to important mathematics content and mathematical practices using your grade-level content standards?	Yes—The task is challenging and will definitely ensure that the students are reasoning abstractly and quantitatively.	Yes—This task is set in a real-world context.	No—This task does not directly match the standard.	Yes—I will have to make sure that I build students' understanding of the equations here before asking them to solve.
How does the task develop, build on, or connect to important mathematical understandings?	The equal sign is placed in different positions in the equation and will prompt important conversations.	Through a context, the students will build understanding of equality.	This is a procedural fluency task.	Students explore how two equivalent equations can be represented.
Does the task provide multiple solution pathways for your students?	Yes—Students can use multiple solution pathways and then prove their answer with counters or sketches.	Yes—There are a few solution pathways, but not as many as for Task A or Task D.	Students are likely to just add. The task does not invite multiple solution pathways.	Yes—There are many solution pathways. Students could strategically select manipulatives or other tools.
Does the task engage your students in *doing mathematics*?	Yes—Students need to represent their understanding with materials or drawings.	Yes—Students are making sense of problems and selecting strategies.	No—This task is not a *doing-math task*. Students will merely apply a procedure.	Yes—The students will demonstrate their understanding of the equal sign.
Does the task require higher-level thinking and reasoning?	Yes—Students can think about the patterns of equations and use manipulatives or draw pictures.	Yes—Students are interpreting the structure of a multistep word problem.	No—These are procedures without connections.	Yes—Students can represent the equations with materials and find the unknown value.
Does the task connect to additional grade-level mathematics topics (or other content areas), and if so, how?	No	No	Yes—The task could connect to probability, but that is not a K–1 topic.	No

Given the analysis of all the information, Carlis immediately ruled out Task C, as she knew that it was not going to challenge her students to think mathematically. Although she loved Task D, she was worried that students needed more experience understanding the meaning of the equal sign. Carlis really liked Tasks A and B. She liked Task A because she thought the students would benefit from proving their ideas with manipulatives or sketches. She liked Task B because she thought that the students would connect with the context because several of the students shared their collections with classmates.

As Carlis reviewed her task selection, she decided to select Task A to help her students in determining if addition and subtraction equations are true or false. She knew that the position of the equal sign would spark a lively conversation with the students and wanted to support them to engage in conversations about the meaning of the equal sign.

Finally, Carlis decided to review her adaptation of Task A with particular consideration of the four special callouts used throughout this book as a lens:

- Access and Equity: Carlis decided to introduce the equations using the Which One Doesn't Belong? lesson launch to encourage students to notice the position of the equal sign.

- Productive Struggle: Carlis's decision to rule out Task C was the first step in promoting productive struggle. She knew that she wanted an open-ended task that would invite students to develop multiple solution pathways, and Task C did not meet that criterion.

- Alternate Learning Environment: Although Task A is not specifically designed for an alternate learning environment, Carlis thought that students would enjoy working on this task at home by constructing an example with materials or sketches for one of the equations posed. She envisioned that students could share ideas during a brainstorming phase through an online discussion board and upload sketches of their ideas or pictures of commonly found items as manipulatives.

- Strengths Spotting: As in any lesson or task, there are many opportunities to spot strengths. Carlis planned to highlight her students' collaboration, strategies, solution pathways, and reasoning throughout the Launch, Facilitate, and Close portions of the lesson.

Your Turn: Adapting Tasks

As you consider adapting a math task, the goal is to transform a task into one that is truly a *doing-math task* and meets your students' instructional needs. One of the main reasons for adapting a task is because the task quality is low, only requiring students to memorize or implement procedures without connecting those procedures to key conceptual understanding of the mathematics they are learning. Another key reason that teachers might want to adapt a task is because the context of the task is not relatable or realistic for their students. There are several ways to adapt math tasks.

ADAPTING A LOW-QUALITY TASK

You can adapt a low-quality task by making changes that provide more opportunities for students to apply prior knowledge, make connections to conceptual understanding, provide multiple solution pathways, explain their thinking, and justify their reasoning. Let's take a look at how a team of kindergarten and first-grade teachers adapted a low-quality place value task to be more challenging and engaging. On the left is the original task, and on the right is the adaptation (Figure 17.3).

Figure 17.3

Order the following numbers:

21, 35, 12, 53

→

Some friends collected baseball cards:

Ray 21 cards

Kalyn 35 cards

Haden 12 cards

Amaya 53 cards

Each friend arranged their cards into groups of 10 and some more. Kalyn said that she had the most groups of 10. Prove how many groups of 10 each friend has to find out who is right.

Another reason a teacher might adapt a task is to strategically design contexts that incorporate students' experiences and funds of knowledge to promote student access. By changing the problem setting to a more familiar and relevant context, students can focus on exploring and learning the mathematics. Teachers can do this by changing the task context using students' shared experiences such as a common field trip they take in a prior grade or are preparing for, community events such as a farmer's market or neighborhood yard sale, or local landmarks or businesses that are explicitly linked to the mathematics that students are learning. Let's take a look at how the kindergarten and first-grade team changed the context of the problem to promote their students' access. The teacher thought the original task was, well, boring and wouldn't inspire students to want to solve the problem. The community arts festival was coming up, however, and that seemed to be a perfect context for the problem. Again, the left task is the original and the right task is the adaptation.

Figure 17.4

Cookies Sold

Chocolate: 23

Peanut Butter: 13

Sugar: 31

Chocolate Chip: 57

What was the most popular? Least?

→

Marian, the baker, was planning to sell some of her famous cookies at the Art Festival. Last year she sold these cookies:

Chocolate: 23

Peanut Butter: 13

Sugar: 31

Chocolate Chip: 57

Which cookie did she sell the most? Least? This year she wants to make 10 more of each kind. How many of each kind will she need?

Finally, you may wish to adapt a task by engaging the students in a task preparation activity to ensure that all students can move quickly into actually *doing the math* that the task requires. This groundwork can include providing a quick review of geometry vocabulary for a task that involves composing two-dimensional shapes to create a composite shape that may involve rectangles, squares, trapezoids, triangles, half-circles, and quarter-circles. It also involves presenting and discussing the use of tools that the students will be expected to use as they engage in the task. You may also consider adapting a task in such a way that small groups of students may be better able to work collaboratively on the task. Such an adaptation may include the use of a group response form designed in such a way that all students provide "evidence" of their involvement in the task's solution strategies.

As these examples demonstrate, there are many ways to adapt a task to promote student engagement, interest, and mathematical understanding. In each instance, you will want to ensure that the task adaptations will support your students' access to the task while maintaining the cognitive rigor.

Your Turn: Creating Tasks

There is no end to the creative possibilities that emerge from a teacher's mind! The more experience teachers have facilitating *doing-math tasks*, the more comfortable they become with creating their own tasks. Use the template we have provided in Appendix A to guide your task construction. Allow us to provide a few recommendations for creating your own *doing-math tasks*:

1. Start with the content standard and the mathematical practices used in your setting as your guideposts. Consistently check the task against these standards to ensure that they are being addressed.

2. Determine your students' learning strengths and challenges. Make your own list, and make sure that the task adequately addresses what students do well, what they are learning, and particular challenges they may be experiencing.

3. Consider your students' interests, experiences, and funds of knowledge. What contexts would interest your students in the task?

As you begin the process of creating your own *doing-math tasks* or adapting tasks from textbooks, online locations, or your school or school district's curriculum resources, find the time to try your "new" tasks out with your colleagues first. Such peer reviews at the pre-implementation stage are always helpful. As tasks are successfully implemented, consider creating shared online files of all tasks so that they can be adapted and revised, as needed, and used by your team in the future.

Finally, as you think about the actual use of a task that you have located, adapted, or created from curriculum materials or related resources, recognize that as you actually implement the task, things change! You may decide to suggest a different representational tool to be used, you may decide to adjust the task to address particular student needs, you may decide to have the class start the task today and finish it tonight on their own, or you may provide time for task completion tomorrow. Such adaptation and flexible implementation decisions are based on the needs and developmental stages of your students and your own instructional awareness and expertise.

As you consider selecting and adapting tasks, we suggest that you begin by starting with the tasks in this book as they are written. Notice and then document how you select and modify the tasks during your planning, or the actual implementation of the task. Next, seek tasks that have similar characteristics and adapt those that need tweaking by using the task selection decision points. Finally, create your own tasks! Go for it!

Frequently Asked Questions

In this work, some questions surface more often than others. Here, we address questions we frequently receive.

Q 1. How can I use the tasks in alternate learning environments (e.g., online, off-site, high level of family support)?

A *Now that many schools are facing a variety of ways to implement instruction, these tasks can fulfill multiple roles. Of course, they can be used in face-to-face settings, but also many of the tasks can be adapted for use in alternate learning environments as noted throughout the task chapters. Consider assigning a task to students in a Google slide or through Jamboard. Multiple students can collaborate on the task by typing or drawing directly on the slide and uploading pictures. Students can also bring family members into the solution process and have them contribute their ideas. We encourage you to adapt tasks to meet these situations in your own creative ways.*

Q 2. How can I monitor student progress using the formative assessment techniques presented within the tasks?

A *Formative assessment tools have been provided for you in Appendix B. Download the tools and adapt them, if needed, to use with the tasks provided in this book or any of the tasks you adapt or create.*

Q 3. What are some ways that I can provide feedback to my students regarding their task performance?

A *We agree that feedback is critical for students to move learning forward. Use the formative assessment tools to record your notes from observations, student interviews, and Show Me responses as well as notes about the student strategies observed and heard and the feedback you provided.*

Q 4. I notice that the tasks provided don't address *all* of the mathematics standards for a grade level. Why is that?

A *These tasks are meant to be a catalyst and to empower you as you are selecting, imagining, and creating additional tasks that meet all of your standards. As much as we wanted to provide tasks that address all of the standards, we simply did not have enough room in this book! We selected those standards that are most often emphasized or highlighted as essential standards for kindergarten and first grade. These are meant to help you get started and empower you to select, adapt, or create additional tasks that meet all of the standards you will teach.*

Summing Up

You began this work on developing a collection of high-quality lessons by considering the important characteristics of a *doing-math task*. By now, we hope that you have had many opportunities to see your students become energized as they engage collaboratively in rich mathematical discourse while developing strategies and solution pathways for the tasks. You have also likely made strategic decisions about the selection and adaptation of these tasks as well as others you will create on your own, always with your students' mathematical strengths at the forefront of your decision making. You have likely developed some specific ideas about what works well with *your* students in *your* classroom, and frankly, this is what this work is all about! Hopefully, you may have also noticed that this process helped solidify your own purpose and intentions for teaching *doing-math tasks*.

As you continue this journey, we suggest the following:

- Find collaboration partners. Seek teachers in your school, across your district, or even across the country to co-create and share task-related ideas. This process is often more rewarding and fun when shared with others.

- Determine how you will document and store your task selection, creation, and adaptation decisions. If you collaborate with others, consider virtual sharing platforms such as Google Docs.

- Celebrate your success by showcasing your students' thinking and reasoning. Make sure that you share these success stories with your leadership!

Finally, as we began to think about the need for this book and the importance of engaging students in *doing-math tasks*, we felt that a resource that helped to define the importance of such tasks and provide exemplars would both assist and stimulate practitioners like you to take the next step and "own" the task creation, adaptation, and implementation process. You see, now, as Smith and Stein (1998) suggested, you surely "know a good task when you see one" (p. 347). Go for it!

Professional Learning/Discussion Questions

- Provide an example of a time when you adapted a math task on the spot during a lesson. What did you do?

- Do your students have any particularly important representation tools that they use regularly as they engage in *doing-math tasks*? What are they?

- What are some particularly helpful sources for mathematics tasks that seem to address, for the most part, this book's task selection decision points?

- How do you, or will you, use formative assessment to monitor student progress as you implement *doing-math tasks*?

- What are some of the challenges you have faced when preparing for and implementing *doing-math tasks* online/virtually?

Appendix A

TASK LESSON TEMPLATE

*The task lesson template provided is the template used for the tasks in Chapters 4–16.

Grade Level

Task Title

Task Topic

Mathematics Standard(s):
Mathematical Practice(s):
Task

Vocabulary	**Materials**

Task Preparation:
Launch:
Facilitate:
Close: Make the Math Visible
Post-Task Notes: Reflection & Next Steps

Appendix B

* Tools are provided for recording student responses to task activities that involve the following formative assessment techniques: Observations, Interviews (individual and small group), and Show Me.

Observations: Small Group or Class

Intent of the Observation	What Was Observed?	Observation Comments; Next Steps
Mathematics Content	(Indicate the mathematics of the task activity)	(What were the students doing math-wise? Quick comment about next steps)
Mathematical Practices	(What processes/practices were students engaged in?)	(What were the students doing with regard to the processes/practices? Next steps?)
Student Engagement	(How were the students engaged in doing the math? Consider: tools and representations grouping: pairs; small groups; communication: discussions; sharing activities; notice and wonder; etc.)	(How was the level of engagement? Did the task truly engage the students in doing math? Next steps?)
General Comment: (Overall comments about the task activities and student involvement in the activities. Particular thoughts about individual, group, or class needs, as observed.)		
Feedback to Students: (How and when will you provide feedback to the students based on your observations of their task performance? This may include interviewing some students or using the Show Me technique based on what you have observed.)		

Source: Adapted from Fennell, Francis (Skip), Kobett, B., & Wray, J. (2017). *The Formative 5: Everyday Assessment Techniques for Every Math Classroom.* Thousand Oaks, CA: Corwin. permission conveyed through Copyright Clearance Center, Inc.

Interview—Individual Student

Student Name:	Date:	Mathematics Topic:
Student Questions:		**Student Responses:** (Provide notes regarding student responses for each of the questions)
1. How did you solve that?		
2. Why did you solve the problem/task that way?		
3. What else can you tell me about what you did?		

Note: If available, attach completed work sample(s).

Source: Adapted from Fennell, Francis (Skip), Kobett, B., & Wray, J. (2017). *The Formative 5: Everyday Assessment Techniques for Every Math Classroom.* Thousand Oaks, CA: Corwin. permission conveyed through Copyright Clearance Center, Inc.

Interviews—Small Group

Student Name	Mathematics Content Focus	Mathematical Practice(s)	Task	Interview Question #1 *(For example: How did you solve that problem?)*	Interview Question #2 *(For example: Why did you solve the problem that way?)*

Note: Add more rows as needed, for additional students.

Source: Adapted from Fennell, Francis (Skip), Kobett, B., & Wray, J. (2017). *The Formative 5: Everyday Assessment Techniques for Every Math Classroom.* Thousand Oaks, CA: Corwin. permission conveyed through Copyright Clearance Center, Inc.

Show Me

Mathematics Content/Standard:	
Task Focus: (title and/or brief description of the content)	**Anticipated Student Show Me Responses:**
Student: (Brief description of the student's Show Me response and/or a picture of the response)	Student:
Student:	Student:

Source: Adapted from Fennell, Francis (Skip), Kobett, B., & Wray, J. (2017). *The Formative 5: Everyday Assessment Techniques for Every Math Classroom.* Thousand Oaks, CA: Corwin. permission conveyed through Copyright Clearance Center, Inc.

References

Adams, T., Thangata, F., & King, C. (2005). "Weigh" to go! Exploring mathematical language. *Mathematics Teaching in the Middle School, 10*(9), 444–448.

Aguirre, J., Mayfield-Ingram, K., & Martin, D. (2013). *The impact of identity in K–8 mathematics: Rethinking equity-based practices.* NCTM.

Anderson, N., Chapin, S., & O'Connor, C. (2011). *Classroom discussions: Seeing math discourse in action, grades K–6.* Math Solutions.

Baker, K., Jessup, N. A., Jacobs, V. R., Empson, S. B., & Case, J. (2020). Productive struggle in action. *Mathematics Teacher: Learning and Teaching PK–12, 113*(5), 361–367.

Ball, D. L. (1992). Magical hopes: Manipulatives and the reform of math education. *American Educator: The Professional Journal of the American Federation of Teachers, 16*(2), 14–18, 46–47.

Ball, D. L., & Forzani, F. M. (2011a). Building a common core for learning to teach, and connecting professional learning to practice. *American Educator, 35*(2), 17–21.

Ball, D. L., & Forzani, F. M. (2011b). *Identifying high-leverage practices for teacher education.* Panel paper presented at the State Higher Education Executive Officer Association Conference, Chapel Hill, NC, May.

Barousa, M. (2013). *Which one doesn't belong?* https://wodb.ca/

Bay-Williams, J. M., & Kling, G. (2019). *Math fact fluency: 60+ games and assessment tools to support learning and retention.* ASCD and NCTM.

Bay-Williams, J. M., & Livers, S. (2009). Supporting math vocabulary acquisition. *Teaching Children Mathematics, 16*(4), 238–246.

Bloom, B. S. (1956). *Taxonomy of educational objectives, handbook I: The cognitive domain.* David McKay.

Boaler, J. (2006). How a detracked mathematics approach promoted respect, responsibility, and high achievement. *Theory Into Practice, 45*(1), 40–46.

Boaler, J. (2014). *Setting up positive norms in math class.* Stanford Graduate School of Education. www.youcubed.org/wp-content/uploads/Positive-Classroom-Norms2.pdf

Boaler, J., & Staples, M. (2008). Creating mathematical futures through an equitable teaching approach: The case of Railside School. *Teachers College Record, 110*(3), 608–645.

Burns, M., & Sheffield, S. (2004). *Math and literature: Grades K–1.* Math Solutions.

Carbonneau, K. J., Marley, S. C., & Selig, J. P. (2013). A meta-analysis of the efficacy of teaching mathematics with concrete manipulatives. *Journal of Educational Psychology, 105*, 380–400.

Cavanaugh, R. A., Heward, W. L., & Donelson, F. (1996). Effects of response cards during lesson closure on the academic performance of secondary students in an earth science course. *Journal of Applied Behavior Analysis, 29*(3), 403–406.

Chapin, S. H., & O'Connor, M. C. (2012). Project Challenge: Using challenging curriculum and mathematical discourse to help all students learn. In C. Dudley-Marling & S. Michaels (Eds.), *High-expectation curricula: Helping all students succeed with powerful learning* (pp. 113–127). Teachers College Press.

Chapin, S. H., O'Connor, C., & Anderson, N. A. (2013). *Talk moves: A teacher's guide for using classroom discussions in math.* Math Solutions.

Danielson, C. (2008). *The handbook for enhancing professional practice: Using the framework for teaching in your school*. ASCD.

Danielson, C. (2016). *Which one doesn't belong? A shapes book and teacher's guide*. Stenhouse.

Danielson, C., & McGreal, T. L. (2000). *Teacher evaluation to enhance professional practice*. ASCD.

Dewey, J. (1933). *How we think: A restatement of the relation of reflective thinking to the educative process*. D. C. Heath.

Dougherty, B., Bryant, D. P., Bryant, B. R., & Shin, M. (2016). Helping students with mathematics difficulties understand ratios and proportions. *Teaching Exceptional Children, 49*(2), 96–105.

Doyle, W. (1988). Work in mathematics classes: The context of students' thinking during instruction. *Educational Psychologist, 23*(2), 167–180.

Drake, C., Land, T. J., Bartell, T. G., Aguirre, J. M., Foote, M. Q., Roth McDuffie, A., & Turner, E. E. (2015). Three strategies for opening curriculum spaces. *Teaching Children Mathematics, 21*(6), 346–353.

Ellis, M. W. (2008). Leaving no child behind yet allowing none too far ahead: Ensuring (in)equity in mathematics education through the science of measurement and instruction. *Teachers College Record, 110*(6), 1330–1356.

Fennell, F., Kobett, B. M., & Wray, J. (2015). Classroom-based formative assessments: Guiding teaching and learning. In C. Suurtamm and A. Roth McDuffie (Eds.), *Annual perspectives in mathematics education—2015* (pp. 51–62). NCTM.

Fennell, F., Kobett, B. M., & Wray, J. (2017). *The formative 5: Everyday assessment techniques for every math classroom*. Corwin.

Fennell, F., & Rowan, T. (2001). Representation: An important process for teaching and learning mathematics. *Teaching Children Mathematics, 7*(5), 288–292.

Flynn, M. (2017). *Beyond answers: Exploring mathematical practices with young children*. Stenhouse.

Franke, M. L., Webb, N. M., Chan, A. G., Ing, M., Freund, D., & Battey, D. (2009). Teacher questioning to elicit students' mathematical thinking in elementary school classrooms. *Journal of Teacher Education, 60*(4), 380–392.

Furner, J. M. (2018). Using children's literature to teach mathematics: An effective vehicle in a STEM world. *European Journal of STEM Education, 3*(3), 14. https://doi.org/10.20897/ejsteme/3874

Fyfe, E. R., McNeil, N. M., Son, J. Y., & Goldstone, R. L. (2014). Concreteness fading in mathematics and science instruction: A systemic review. *Educational Psychology Review, 26*, 9–25.

Ganske, K. (2017). Lesson closure: An important piece of the student learning puzzle. *The Reading Teachers, 71*(1), 95–100.

Gonzalez, J. (2018, December 2). 10 ways educators can take action in pursuit of equity [Blog]. *Cult of Pedagogy*. https://www.cultofpedagogy.com/10-equity/

Hattie, J. (2009). *Visible Learning: A synthesis of over 800 meta-analyses relating to achievement*. Routledge.

Heddens, J. W. (1986). Bridging the gap between the concrete and the abstract. *Arithmetic Teacher, 33*(6), 14–17.

Heick, T. (2019). 10 ways to be a more reflective teacher [Blog post]. *Teachthought*. https://www.teachthought.com/pedagogy/reflective-teacher-reflective-teaching/

Hiebert, J., Carpenter, T. P., Fennema, E., Fuson, K., Human, P., Murray, H., Olivier, A., & Wearne, D. (1996). Problem solving as a basis for reform in curriculum and instruction: The case of mathematics. *Educational Researcher, 25*(4), 12–21.

Hiebert, J., & Wearne, D. (1993). Instructional tasks, classroom discourse, and students' learning in second-grade arithmetic. *American Educational Research Journal, 30*(2), 393–425.

Illustrative Math. (2019). Co-creating classroom norms with students [Blog]. https://illustrativemathematics.blog/2019/08/02/co-creating-classroom-norms-with-students/

Jackson, K. J., Shahan, E. C., Gibbons, L. K., & Cobb, P. A. (2012). Launching complex tasks. *Mathematics Teaching in the Middle School, 18*(1), 24–29.

Jilk, L. M. (2016). Supporting teacher noticing of students' mathematical strengths. *Mathematics Teacher Educator, 4*(2), 188–199.

Kapur, M. (2010). Productive failure in mathematical problem solving. *Instructional Science, 38*(6), 523–550.

Karp, K. S., Dougherty, B. J., & Bush, S. B. (2020). *The math pact, elementary: Achieving instructional cohesion within and across grades.* Corwin and NCTM.

Kasberg, S. E., & Frye, R. S. (2013). Norms and mathematical proficiency. *Teaching Children Mathematics, 20*(1), 28–35.

Kaye, P. (2012). *Games for math: Playful ways to help your child learn math: From kindergarten to third grade.* Pantheon.

Kazemi, E., & Hintz, A. (2014). *Intentional talk: How to structure and lead productive mathematical discussions.* Stenhouse.

Keeley, P., & Tobey, C. R. (2011). *Mathematics formative assessment, volume 1: 75 practical strategies for linking assessment, instruction, and learning.* Corwin.

Kelemanik, G., & Lucenta, A. (2016). *Routines for reasoning: Fostering the mathematical practices in all students.* Heinemann.

Kobett, B. M., Harbin Miles, R., & Williams, L. (2018). *The mathematics lesson-planning handbook, grades K–2: Your blueprint for building cohesive lessons.* Corwin.

Kobett, B., & Karp, K. (2020). *Strengths-based teaching and learning in mathematics: 5 teaching turnarounds for grades K–6.* Corwin and NCTM.

Laski, E. V., Jor'dan, J. R., Daoust, C., & Murray, A. K. (2015). What makes mathematics manipulatives effective? Lessons from cognitive science and Montessori education. *SAGE Open, 5*(2), 1–8.

Lempp, J. (2017). *Math workshop: Five steps to implementing guided math, learning stations, reflection, and more.* Math Solutions.

Livers, S., & Bay-Williams, J. M. (2014). Vocabulary support: Constructing (not obstructing) meaning. *Mathematics Teaching in the Middle School, 20*(3), 152–159.

Lortie, D. C. (1975). *Schoolteacher: A sociological study.* University of Chicago Press.

Lynch, S. D., Hunt, J. H., & Lewis, K. E. (2018). Productive struggle for all: Differentiated instruction. *Mathematics Teaching in the Middle School, 23*(4), 194–201.

Marshall, A. M., Superfine, A. C., & Canty, R. (2010). Star students make connections: Discover strategies to engage young math students in competently using multiple representations. *Teaching Children Mathematics, 17*(1), 38–47.

Math Forum. (2015). Beginning to problem solve with "I notice, I wonder." mathforum.org/pow/noticewonder/intro.pdf

Murata, A., & Stewart, C. (2017). Facilitating mathematical practices through visual representations. *Teaching Children Mathematics, 23*(7), 404–412.

Murphy, S. J. (2000). Children's books about math: Trade books that teach. *New Advocate, 13*(4), 365–374.

National Council of Teachers of Mathematics. (1991). *Professional standard for teaching mathematics.* Reston, VA: National Council of Teachers of Mathematics.

National Council of Teachers of Mathematics. (2000). *Principles and standards for school mathematics*. Author.

National Council of Teachers of Mathematics. (2014a). *Access and equity in mathematics education*. Author.

National Council of Teachers of Mathematics. (2014b). *Principles to actions: Ensuring mathematical success for all*. Author.

National Council of Teachers of Mathematics. (2017a). *Exploring math through literature, pre-K–8*. Author. https://www.nctm.org/Store/Products/Exploring-Math-through-Literature-Pre-K-8/

National Council of Teachers of Mathematics. (2017b). *Principles to actions: Professional learning toolkit*. Author. https://www.nctm.org/PtAToolkit/

National Council of Teachers of Mathematics. (2017c). *Taking action: Implementing effective mathematics teaching practices in K–grade 5*. Author.

National Council of Teachers of Mathematics. (2020a). *Catalyzing change in early childhood and elementary mathematics: Initiating critical conversations*. Author.

National Council of Teachers of Mathematics. (2020b). *Classroom practices that support equity-based mathematics teaching*. Research brief. https://www.nctm.org/Research-and-Advocacy/Research-Brief-and-Clips/Classroom-Practices-That-Support-Equity-Based-Mathematics-Teaching/

National Council of Teachers of Mathematics. (n.d.). Illuminations. https://illuminations.nctm.org/

National Governors Association Center for Best Practices & Council of Chief State School Officers. (2010). *Common Core state standards for mathematics*. Author.

National Research Council. (2001). *Adding it up: Helping children learn mathematics*. National Academies Press.

National Research Council. (2012). *Education for life and work: Developing transferable knowledge and skills in the 21st century*. Committee on Defining Deeper Catalyzing Change in High School Mathematics.

O'Connell, S., & SanGiovanni, J. (2013). *Putting the practices into action: Implementing the Common Core standards for mathematical practice, K–8*. Heinemann.

Opfer, V. D., Kaufman, J. H., & Thompson, L. E. (2017). *Implementation of K–12 state standards for mathematics and English language arts and literacy: Findings from the American teacher panel*. RAND.

Parrish, S. (2014). *Number talks: Whole number computation, Grades K–5*. Math Solutions.

Pollock, J. E. (2007). *Improving student learning one teacher at a time*. ASCD.

Raposo, J., & Stone, J. (1972). One of these things is not like the other [Song lyrics]. www.metrolyrics.com/one-of-these-things-is-not-like-the-others-lyrics-sesame-street.html

Ritchhart, R., Church, M., & Morrison, K. (2011). *Making thinking visible: How to promote engagement, understanding, and independence for all learners*. Wiley.

Rumack, A., & Huinker, D. (2019). Capturing mathematical curiosity with notice and wonder. *Mathematics Teaching in the Middle School, 24*(7), 394–399.

SanGiovanni, J. J., Katt, S., & Dykema, K. (2020). *Productive math struggle: A 6-point action plan for fostering perseverance*. Corwin.

Shor, C. (2017). Clothesline math. https://clotheslinemath.com

Silver, E. A., & Mills, V. L. (Eds.). (2018). *A fresh look at formative assessment in mathematics teaching*. NCTM.

Small, M. (2017). *Good questions: Great ways to differentiate math instruction in standards-based classrooms*. Teacher's College Press.

Smith, M. S., Bill, V., & Sherin, M. G. (2019). *The 5 practices in practice: Successfully orchestrating mathematics discussions in your elementary classroom*. Corwin.

Smith, M. S., & Stein, M. K. (1998). Selecting and creating mathematical tasks: From research to practice. *Mathematics Teaching in the Middle School, 3*(5), 344–349.

Smith, M. S., & Stein, M. K. (2018). *Five practices for orchestrating productive mathematics discussions* (2nd ed.). NCTM.

Smith, N. N. (2017). *Every math learner, grades K–5: A doable approach to teaching with learning differences in mind*. Corwin.

Stein, M. K., Grover, B. W., & Henningsen, M. (1996). Building student capacity for mathematical thinking and reasoning: An analysis of mathematical tasks used in reform classrooms. *American Educational Research Journal, 33*(2), 455–488.

Stein, M. K., & Lane, S. (1996). Instructional tasks and the development of student capacity to think and reason: An analysis of the relationship between teaching and learning in a reform mathematics project. *Educational Research and Evaluation, 2*(1), 50–80.

Stein, M. K., Smith, M. S., Henningsen, M., & Silver, E. A. (Eds.). (2000). *Implementing standards-based mathematics instruction: A casebook for professional development*. Teachers College Press.

Stein, M. K., Smith, M. S., Henningsen, M. A., & Silver, E. A. (Eds.). (2009). *Implementing standards-based mathematics instruction: A casebook for professional development* (2nd ed.). Teachers College Press.

Stigler, J. W., & Hiebert, J. (2004). Improving mathematics teaching. *Educational Leadership, 61*(5), 12–16.

Sweller, J. (1988). Cognitive load during problem solving: Effects on learning. *Cognitive Science, 12*(2), 257–285.

Thompson, D. R., & Rubenstein, R. N. (2000). Learning mathematics vocabulary: Potential pitfalls and instructional strategies. *The Mathematics Teacher, 93*(7), 568–574.

Turner, E., Dominguez, H., Maldonado, L., & Empson, S. (2013). English learners' participation in mathematical discussion: Shifting positionings and dynamic identities. *Journal for Research in Mathematics Education, 44*(1), 199–234.

Van de Walle, J., Karp, K., & Bay-Williams, J. (2019). *Elementary and middle school mathematics: Teaching developmentally*. Pearson.

Van de Walle, J. A., Karp, K. S., Lovin, L. H., & Bay-Williams, J. M. (2017). *Teaching student-centered mathematics: Developmentally appropriate instruction for grades 3–5* (3rd ed., Vol. 2). Pearson.

White, D. Y., Gomez, C. N., Rushing, F., Hussain, N., Patel, K., & Pratt, J. (2018). Assembling the puzzle of mathematical strengths. *Mathematics Teaching in the Middle School, 23*(5), 268–275.

Wiliam, D. (2011). *Embedded formative assessment*. Solution Tree Press.

Wiliam, D., & Thompson, M. (2008). Integrating assessment with learning: What will it take to make it work? In C. A. Dwyer (Ed.), *The future of assessment: Shaping teaching and learning* (pp. 52–82). Lawrence Erlbaum.

Willingham, D. T. (2017). Ask the cognitive scientist: Do manipulatives help students learn? *American Educator*. https://www.aft.org/ae/fall2017/willingham

Yeh, C. (2019). *Countering deficit myths of students with dis/abilities and conceptualizing possibilities: A culturally responsive and relational approach to mathematics TODOS Live!* [Webinar] https://vimeo.com/353856573

Yeh, C., Ellis, M. W., & Hurtado, C. K. (2017). *Reimagining the mathematics classroom: Creating and sustaining productive learning environments*. NCTM.

Zaslavsky, C. (1998). *Math games & activities from around the world*. Chicago Review Press.

Index

A SAGE Publishing Company

Helping educators make the greatest impact

CORWIN HAS ONE MISSION: to enhance education through intentional professional learning.

We build long-term relationships with our authors, educators, clients, and associations who partner with us to develop and continuously improve the best evidence-based practices that establish and support lifelong learning.

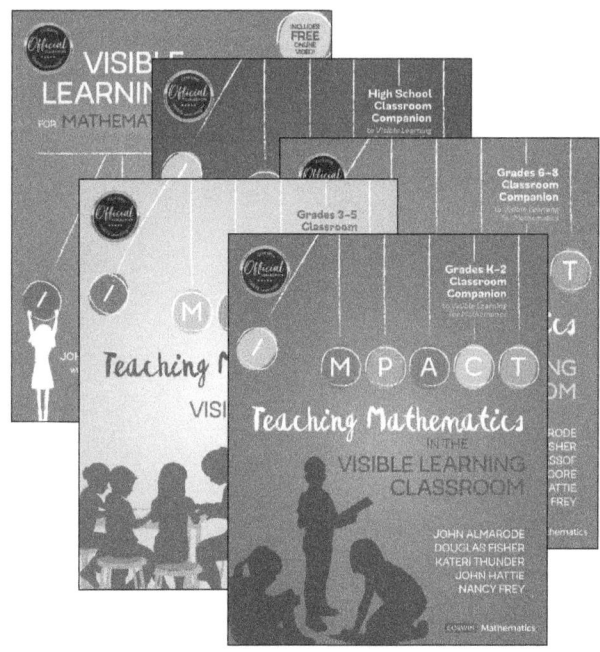
CMN21129

CORWIN